Transferable Groundw:

The principle of transferable groundwater rights is that by making water rights capable of being traded in the market, water resources can be used more sustainably and efficiently. Groundwater would achieve its economic value, by switching from the high-volume, low-value irrigation, which is prevalent with many farmers, particularly in South Asia, to low-volume, high-value urban supply or the growing of intensive horticultural or cash crops.

This book discusses transferable groundwater rights in their broader context. It starts with a detailed description of the physical aspects of groundwater, which non-technical readers should find useful, followed by a discussion of legal and economic aspects. Water transfers and the international experiences in transferable groundwater rights are dealt with in detail in two subsequent chapters. A model is presented to guide those involved in water resources management and planning in their decision process to introduce transferable groundwater rights and water rights trading.

The author concludes that transferable groundwater rights potentially offer a better alternative to land-based water rights systems. However, he casts serious doubt on whether groundwater rights trading on its own can achieve water resources sustainability, environmental protection and social equity. Government intervention seems to be almost always needed to assist the water rights market and take responsibility for any of its adverse consequences.

Andreas N. Charalambous is a Consultant and Director of Hydrolaw Ltd, Guildford, UK. He has postgraduate qualifications in hydrogeology from University College, London, and in water law from the UNESCO Centre for Water Law and Policy, University of Dundee.

Earthscan Studies in Water Resource Management

Water Management, Food Security and Sustainable Agriculture in Developing Economies
Edited by M. Dinesh Kumar, M.V.K. Sivamohan and Nitin Bassi

Governing International Watercourses
River basin organizations and the sustainable governance of internationally shared rivers and lakes
Susanne Schmeier

Transferable Groundwater Rights
Integrating hydrogeology, law and economics
Andreas N. Charalambous

Contemporary Water Governance in the Global South
Scarcity, marketization and participation
Edited by Leila Harris, Jacqueline Goldin and Christopher Sneddon

For more information and to view forthcoming titles in this series, please visit the Routledge website: http://www.routledge.com/books/series/ECWRM/

Transferable Groundwater Rights

Integrating hydrogeology, law and economics

Andreas N. Charalambous

First edition published 2013
by Routledge
2 Park Square, Milton Park, Abingdon, Oxon OX14 4RN

Simultaneously published in the USA and Canada
by Routledge
711 Third Avenue, New York, NY 10017

Routledge is an imprint of the Taylor & Francis Group, an informa business

First issued in paperback 2016

British Library Cataloguing in Publication Data
A catalogue record for this book is available from the British Library

Library of Congress Cataloging-in-Publication Data
Charalambous, Andreas N.
    Transferable groundwater rights : integrating hydrogeology, law
    and economics / Andreas N. Charalambous. – 1st edition.
        pages cm. – (Earthscan studies in water resource
        management)
    Originally presented as the author's thesis (LLM)–University of
    Dundee, Scotland.
    Includes bibliographical references and index.
    1. Water rights. 2. Groundwater–Law and legislation. I. Title.
    K3496.C48 2013
    333.33′9–dc23   2012041706

ISBN: 978-0-415-50724-0 (hbk)
ISBN: 978-1-138-68029-6 (pbk)

Typeset in Baskerville
by HWA Text and Data Management, London

This book is dedicated to the memory of my father Nico and mother Poulia

Contents

Tables

Figures

Preface

Transferable groundwater rights, the subject of this book, came to prominence in the 1980s and 1990s, as part of the then more or less worldwide political and economic shift to the right. The theory was that by making groundwater rights transferable, and therefore capable of being traded in the open market, groundwater resources could be used more sustainably and efficiently. Switching from high-volume, low-value groundwater irrigation, which is prevalent in many parts of South Asia, to low-volume, high-value urban or industrial supply or the growing of cash crops was seen as the mechanism for water achieving its economic value. Sceptics pointed out that free markets do not usually serve well the environment, the vulnerable in society and the conservation of resources. Given the substantial staple food price increases of the last few years, the impact on poor families of small holders who might have been tempted to sell their groundwater rights for short-term gain could have been devastating.

This book arose from my research on transferable water rights for a Master of Laws (LLM) at the University of Dundee, Scotland. The book sets out to discuss transferable groundwater rights in a broad context. Thus, whilst the focus is on legal and economic aspects, there is a fairly comprehensive description of the physical aspects of groundwater, which non-technical readers should find useful. Based on a detailed review of the international experience, the book presents a critical assessment of whether groundwater rights trading is the answer to sustainable groundwater management, or fraught with potential dangers to the environment and social equity. It concludes that transferable groundwater rights potentially offer a better alternative to land-based groundwater rights systems. But it casts doubt on whether groundwater rights trading can achieve water resources sustainability, let alone protect the environment or safeguard the welfare and livelihoods of poor and vulnerable communities.

Andreas Charalambous
Guildford
2012

Acknowledgements

I am grateful to my friend and former colleague Peter Garratt, who has had the patience to read through the entire manuscript. I am also grateful to Tom Hargreaves, a friend for many years and former colleague, for preparing the figures and maps. To my son Michael, my thanks for his help with the mysteries of Word, Excel and the like.

My thanks to my reviewers, Bruce Misstear of Trinity College Dublin, Ireland, Andrew Allan of Dundee University, Scotland, Phoebe Koundouri of Athens University of Economics and Business, Greece and Henning Bjornlund of the Universities of South Australia and Lethbridge, Canada. Also, to two other reviewers for the publishers. Their comments added value to the book. My special thanks to Andrew Allan, who encouraged me to write this book.

My thanks also go to Tim Hardwick of Earthscan/Routledge, for his help and support in the last two years.

Rose, my wife, showed admirable tact in expressing her keen interest and Nicole, my daughter, stayed in the wings. My thanks to both.

Abbreviations

l	litre
l s^{-1}	litres per second
km^3	cubic kilometre
km^3 a^{-1}	cubic kilometres per year
m	metre
mg	milligramme
mg l^{-1}	milligramme per litre
m^3	cubic metre
μS cm^{-1}	microsiemens per centimetre
Mm^3	million cubic metres
Mm^3 a^{-1}	million cubic metres per year
ppm	parts per million

Conversions

1 m^3 = 1000 l
1 km^3 = 1000 Mm^3
1 km^3 = 810,713.194 acre-feet
1 l s^{-1} = 13.198 Imperial gallons per minute
1 l s^{-1} = 15.85 US gallons per minute
1 l s^{-1} = 0.031536 Mm^3 a^{-1}
1 Mm^3 = 810.713194 acre-feet

1 Introduction

Historically water rights legislation focused on surface water. In England, the Industrial Revolution in the eighteenth century saw an increase in the use of water initially to provide energy for water-driven mills, and later for use in industrial processes and agriculture. Competing interests for access to flowing water in rivers and streams resulted in much legal conflict over water rights, and the development of water laws related mainly to riparian possession. In the arid regions of the western USA where water is scarce, riparian laws gave way to the prior appropriation doctrine, which enabled nineteenth-century water users, miners and farmers, to use water away from riparian lands on the basis of a 'first in time, first in right' principle.

Groundwater law has been a little slow to evolve, partly because the flow of water underground was not well understood, and partly because it was plentiful – taken in small quantities mainly from springs that flowed naturally or from shallow hand-dug wells hand-drawn by buckets or with the help of animals. Traditionally, a landowner could extract an unlimited amount of groundwater found beneath his land without concern for his neighbours. This is generally known as the 'absolute ownership doctrine', and stems from both Roman law and the common law of England. It is still the groundwater law in many developing countries and in the state of Texas in the US where it is known as the 'rule of capture'. Changes in legislation started around the middle of the last century, mainly in response to an increase in the use of groundwater, which has been made possible by advances in drilling and pump technology. Nowadays, high-capacity water wells can be constructed to depths of several hundred metres in a couple of months or less, whilst electrically driven submersible pumps can extract groundwater at high rates from greater depths. The adverse effects of over-abstraction are well known: deepening groundwater levels, depletion of storage reserves, deteriorating water quality and reduction in river flows, impacts on wetlands, ground subsidence etc. In order to protect and conserve groundwater resources, modern legislation has sought to introduce formal and explicit water rights that restrict the volume and duration of abstraction; also in order to make groundwater management more flexible, it has sought to break the traditional link between land ownership and groundwater rights and allow water rights to be treated separately from land rights. Generally, the approach to water resources management has been regulation by the state. Deregulation is relatively recent. It started in the 1980s, the child of free market economists, and supported, at least initially, by institutions such as the World Bank. At that time, it was seen to be the answer to the ever-increasing groundwater abstraction for subsidised irrigation in South Asia, and Central and South America.

Transferable groundwater rights (TGR) that can be freely traded are at the heart of the free market philosophy. Legislation to achieve this was introduced in Chile in 1986 by the right-wing junta of General Pinochet. A few other countries (Mexico, Australia, England and Wales) followed suit, though in none was water legislation as 'pure free market' as in Chile; and even in Chile it was soon recognised that the market alone did not have all the answers. But as always, there are lessons to be learned: appreciating that water has an economic value is one of them. Finding ways to prevent wastage and use water more efficiently is another. Both are valuable to developing economies, where the conflicts of ever-increasing city populations and the requirements for farming and social cohesion of rural communities present dilemmas that have no easy solutions.

2 Elements of hydrogeology

The purpose of this chapter is to provide the technical basis for the legal and economic discussions that follow in subsequent chapters. The chapter provides a broad view of hydrogeology, with emphasis on general principles, avoiding, as far as practically possible, detailed technical discussions. It begins by defining hydrogeology and continues with a note on the historic development of water science and hydrogeology from ancient times to the present day. A short discussion follows on the nature of water, the hydrologic cycle and infiltration into aquifers. Groundwater is dealt with at some length. Aspects discussed include: its occurrence and origin, movement and storage underground, recharge and abstraction, chemical quality and pollution, and management in the face of over-abstraction. An effort has been made to present an international view of groundwater, particularly in terms of its exploitation and the impacts of this in recent years. It has not always been possible to avoid technical terms, but where these occur they have been clearly and simply explained, with definitions of geological and hydrogeological terms as appropriate.

What is hydrogeology?

Hydrogeology is concerned with the study of water that occurs below the surface of the earth. The term 'hydrogeology' was probably first used by Lamarck (1802) in his book *Hydrogéologie*, which, however, did not deal with groundwater but rather with the influence of water on the surface of the earth. Probably the first modern use of the term in Britain was by Lucas (1877a), a junior officer in the Geological Survey of Britain, who wrote:

> Hydrogeology … takes up the history of rain-water from the time that it leaves the domain of the meteorologist and investigates the conditions under which it exists in passing through the various rocks which it percolates after leaving the surface.

And more concisely in a further paper (Lucas, 1877b):

> Hydrogeology takes up the history of rain water from the time that it touches the soil and follows it through the various rocks which it subsequently percolates.

About 40 years later, the American engineer Mead in his book *Hydrology* (1919) defined hydrogeology as 'the study of the occurrence and movement of subterranean waters'. A more elaborate definition was given by Davis and DeWiest (1966) who defined hydrogeology as 'the study of groundwater with particular emphasis given to its chemistry, mode of migration and relation to the geologic environment'; and Domenico and Schwartz (1990) as

> the study of the laws governing the movement of subterranean water, the mechanical, chemical, and thermal interaction of this water with the porous solid, and the transport of energy and chemical constituents by flow.

The term 'geohydrology' was used by Davies and DeWiest to describe the hydrological or fluid aspects of groundwater. Todd (1959) used the somewhat cumbersome term, 'groundwater hydrology' in his textbook on groundwater, which he defined as 'the science of the occurrence, distribution and movement of water below the surface of the earth'. The term has been retained in the latest publication (Todd and Mays, 2005) and has also been used by Heath (2004). Earlier, Meinzer (1942), considered the term 'hydrology' to include 'the occurrence of water in the Earth, its physical and chemical reactions with the rest of the Earth, and its relation to the life of the Earth'. In modern usage, however, the term is mainly applied to the study of surface water and surface water processes. However, Hiscock (2005) suggests that as hydrogeology encompasses aspects of hydrology too strict a division between the two is unhelpful.

Ancient ideas and modern science

It is not clear to what extent the ancients understood groundwater, although it seems that they were aware that rainfall infiltrates into rocks and that springs are related to rainfall. But they generally seemed to have been rather confused and felt it necessary to invoke large subterranean lakes to supplement rainfall.

In Greek mythology the source of all rivers, wells and springs was thought to be from Oceanus, a great river encircling the Earth, and personified by a Titan, born to Uranus and Gaia. Thus, in Book 21 of *The Iliad*, Homer (who lived around 850 BC) has Achilles perorating over the dead body of Asteropaeus as follows:

> but it may not be that one should fight with Zeus the son of Cronos. With him doth not even King Achelous vie, nor the great might of deep-flowing Ocean, from whom all rivers flow and every sea, and all the springs and deep wells.

The Greek philosopher Anaxagoras (c. 500–428 BC) thought that rivers owed their origin to rainfall and to water stored in cavities in the hollow earth. Plato (427–347 BC) in *Critias* seems to have had the right idea on the origin of rivers and springs, when he says:

the land reaped the benefit of the annual rainfall, ... and receiving it into herself ... it let off into the hollows the streams which it absorbed from the heights, providing everywhere abundant fountains and rivers.

(translation by Jowett, 1892)

Aristotle, (384–322 BC) the pupil of Plato, in Book 1, Part 13 of his *Meteorologica*, while accepting that rivers have their source in rainfall, also suggests that the condensation of 'air' underground forms another source of river water, with droplets gathering together in a way analogous to water seeping from the soil during the construction of trenches to collect water in pipes or aqueducts. He makes the point that mountains and high ground where large springs and rivers originate act as a sponge absorbing rainfall and also cool the vapour that rises in the rock. Interestingly, he also refers to river waters being swallowed into fractures and cavities in Arcadia of the Peloponnese, and at a larger scale outside Greece, and emerging elsewhere. The concepts of the Greek philosophers of large subterranean spaces where hydrological processes took place were probably influenced by their familiarity with cavernous limestones and springs emerging from them. However, due to the lack of measurement they had difficulty in accepting that water from springs and rivers is derived from rainfall, and had to invent fanciful subterranean sources and processes.

Although the Romans made great advances in tapping, storing and conveying water, in science they largely followed the ideas of the Greeks. One exception was Vitruvius, a Roman architect who lived probably around 70–25 BC. In his *De Architectura* (circa 28 BC) he correctly surmised that rain and snow falling in large quantities on mountains infiltrate through the rocks to appear at the foothills and give rise to springs. He also indicated which plants could be used as indicators of groundwater and explained the process of alkalisation of solids by evaporation of groundwater. Nevertheless, the ideas of the Greeks persisted until the Renaissance, when notably Palissy (1580) presented convincing evidence that the source of springs is from rainwater that seeps into the Earth sinking to impervious rock until it reaches an outlet. However, it was not until much later when Perrault (1678), Mariotte (1686) and Halley (1687) were able to demonstrate by means of systematic measurements and observation that rainfall was more than sufficient to account for river flow.

A history of the development of ideas on the hydraulic characterisation of aquifers, reservoir rocks and oils has been presented by Narasimhan (1998). Fetter (2004) provided a short history of hydrogeology in the last 200 years. Quantitative hydrogeology started with Darcy (1856) whose experimental work gave rise to Darcy's law that relates velocity of flow to permeability and hydraulic gradient. Following from this, Dupuit (1863) and independently Forcheimer (1886) provided regional groundwater flow solutions, which became known as the Dupuit–Forchheimer approximation. Thiem (1887) developed equations to describe the steady flow of groundwater into wells. Theis (1935) used equations analogous to heat conduction in solids – originally proposed by Fourier in 1807 – to describe transient or time-dependent flow. These mathematical solutions formed the basis of subsequent work predicting the position of groundwater levels when groundwater is pumped out and the impact on other users, springs and surface water bodies.

Models to simulate groundwater flow were introduced in the late nineteenth century, initially using sand-tanks, and later Hele-Shaw viscous flow and electric analogue models (Hele-Shaw, 1898; Karplus, 1958; Polubarinova-Kochina, 1962). With the increasing power of computers the earlier physical techniques were mostly replaced by computer modelling (Prickett and Lonnquist, 1971; McDonald and Harbaugh, 1988). Over the last 10–15 years, computer software packages capable of simulating a variety of groundwater conditions have become widely available.

In modern times, measurement constitutes the basis for the evaluation of groundwater resources. Instrumentation has become more sophisticated and reliable, while the development of digital technology and computer software has enabled the recording, storing and processing of large amounts of data. Today there is a reasonably good coverage of groundwater, water resources and water quality monitoring networks in most countries, including climate stations, rain gauges, river flow gauging stations and groundwater level observation boreholes. Government departments have been created responsible for the monitoring, evaluation and management of water resources. Recently, there have been intensive efforts to computerise records and allow public access to them.

The nature and unique properties of water

The water molecule consists of two hydrogen atoms and one oxygen atom in an asymmetric structure, which causes it to have polarity, i.e., one side (the hydrogen side) to be positively charged and the other side (the oxygen side) to be negatively charged. As a result individual water molecules attract each other (hydrogen bond) and, relative to other molecules of similar molecular weight, greater energy is needed to break them up. This is why water, despite its low molecular weight, occurs as a liquid under normal temperatures and pressures. Other unique properties contribute significantly to making life on Earth sustainable. For example, a maximum density of 1.00013 grams per cubic centimetre (g cc^{-1}) at 3.98 degrees Celsius (°C) allows liquid water to occur beneath the less dense ice and thus aquatic life in the higher latitudes to survive when the upper parts of water bodies are frozen. The high specific heat of water in relation to most other substances (1.007 calories per gram per degree Celsius (cal $g^{-1}°C^{-1}$) at 0°C) makes it possible for warm-blooded animals to regulate their temperatures, and for the oceans and other water bodies to act as moderators to the changes of the rates and magnitudes of ambient temperature variations. Also, its very large latent heat of vaporisation (595.9 cal g^{-1}) plays an important role in global latitudinal heat transport, and as a source of energy that drives the precipitation-forming process and as a mechanism for transferring large amounts of heat from the Earth's surface to the atmosphere. Furthermore, its surface tension (75.6 dynes cm^{-1} at 0°C) is higher than most other liquids. In granular aquifers the zone above the water table, known as the capillary fringe, is due to surface tension. Finally water is an excellent solvent, an important property in the process of weathering of rocks and the presence of dissolved substances in water which are essential to life.

The hydrological cycle

The hydrological cycle is one of the most important concepts in the understanding of the natural physical processes of water bodies on the earth and the atmosphere. It describes

> the pathway of water as it moves in its various phases through the atmosphere, to the Earth, over and through the land, to the ocean, and back to the atmosphere.

> (National Research Council, 1991)

The cycle is shown graphically in Figure 2.1.

The main source of water to the Earth is from precipitation (rainfall and snow). Other sources from space or from volcanic venting of water vapours are minor. The hydrological cycle is continuous: water from the oceans and the land surface evaporates and the water vapour is transported through the atmosphere where it cools and condenses to form cloud droplets or ice crystals, which return to Earth as precipitation. Total global precipitation on both land and the oceans is about 505,000 km^3, of which only about 21 per cent falls on land (National Research Council, 1986). A large part, about 60 per cent, of the precipitated water on land evaporates, and the rest, about 40 per cent, runs off directly into streams, rivers and lakes or becomes groundwater, emerging as springs or feeding the

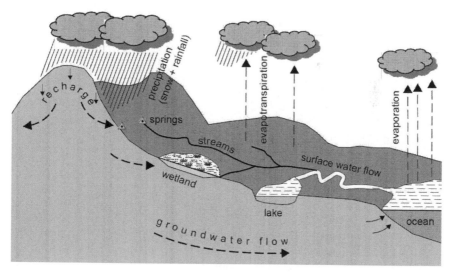

Figure 2.1 Simplified hydrological cycle

baseflow of rivers. The groundwater component amounts to about 30 per cent of the total (surface water and groundwater) renewable water resources (Döll and Fiedler, 2008) or about 12 per cent of land rainfall. Fresh water accounts for only 2.53 per cent of the total world reserves of 1,385,984,610 km^3 and groundwater 0.76 per cent. The bulk of the stored reserves, 96.5 per cent, is in the oceans (Shiklomanov and Sokolov, 1983). However, 96.4 per cent of the unfrozen fresh water is groundwater, with only 3.5 per cent in rivers and lakes.

Groundwater

Other names used in the literature are 'ground water' or 'ground-water'. The modern usage is one word, 'groundwater' without the hyphen or space between the words. The terms 'underground water' and its Latin equivalent 'subterranean water' are normally used to denote water in both the saturated and the unsaturated zones. Meinzer (1968) suggested the term 'subsurface water' for all waters below the surface, in contrast to surface water, which describes water on the surface.

Definitions of groundwater vary slightly but they generally refer to underground water that occupies the saturated zone (UNESCO/WMO, 1992; European Commission, 2000). This is the zone where all the voids in a rock are occupied by water. Soil moisture is not groundwater. Nor is water that flows in well defined underground channels or subterranean streams.

Groundwater is found in all types of rocks. In sands and gravels it occurs in the pores between grains, and in very fine materials, such as clays and silts, in microscopic pores that have little connection between them. In hard basement rocks and lavas, and in sandstones and limestones, groundwater occurs in fractures. In limestones, fractures are often enlarged by the action of water to form caves and large voids.

As already mentioned, until fairly recently the origin of groundwater was a matter of conjecture. It is now accepted that almost all groundwater is of meteoric origin, meaning that it is derived from the infiltration of rainfall or surface water bodies into the subsurface geological strata, and therefore a part of the hydrological cycle. There are some minor additions from igneous activity (magmatic or plutonic waters). Hydrothermal waters that are found in thermal (hot) springs or geysers are also of meteoric origin, and represent surface waters that have travelled to great depths in areas of volcanic activity where earth temperatures are high. Groundwaters which have been part of geological processes are known as 'connate waters' and are usually very old and mineralised.

Non-renewable and fossil groundwater

Most groundwater remains underground for a relatively short time, generally for a few months, and often for a few years, before it finds its way out to the surface into rivers, or as springs and seepages. In the arid areas of the Middle East and the Sahara Desert in North Africa, groundwaters are 20,000–40,000 years old, the relics of much wetter periods. These groundwaters are generally known as 'fossil groundwaters'. Others have distinguished between 'fossil' and 'non-renewable' groundwaters, the former as having received no recharge over the millennia and the latter as receiving a 'very low current rate of average annual renewal' (Foster and Loucks, 2006).

Types of water-bearing strata: aquifers, aquicludes and aquitards

Hydrogeologists have classified water-bearing strata into three types, namely, aquifers, aquicludes and aquitards, according to their ability to store and transmit water. A fourth type, aquifuge, is also often included to represent dense rocks that neither store nor transmit water. The names are derived from Latin. *Aqui* is the root for *aqua* (water); *–fer* is the suffix from *ferre*, which means to bear, and therefore an aquifer is a water bearer. The suffix *–clude* in aquiclude is derived from *claudere* (to shut or close), *–tard* in aquitard from *–tardus* (slow), and *–fuge* in aquifuge from *fugere* (to drive away).

The classification is more descriptive than precise, although each type lies within a certain range of physical properties. In general, aquifers have hydraulic conductivities of 0.1 to more than 10,000 m d^{-1}, aquitards 0.01 to 0.001 m d^{-1} and aquicludes less than 0.0001 m d^{-1} (Table 2.1). However, very thick formations composed of materials with very low hydraulic conductivities (less than 0.1 m d^{-1}) may also qualify as aquifers.

An aquifer has been variously defined as a geological formation that contains sufficient saturated permeable material to yield significant quantities of groundwater to wells and springs (Lohman *et al.*, 1972); or as a permeable water-bearing formation capable of yielding exploitable quantities of groundwater (UNESCO/WMO, 1992); or as a subsurface layer or layers of rock or other geological strata of sufficient porosity and permeability to allow either a significant flow of groundwater or the abstraction of significant quantities of groundwater (European Commission, 2000). The qualification that an aquifer should yield

Table 2.1 Range of typical general values of hydraulic properties of water-bearing strata (values of hydraulic parameters from various sources)

Type of water-bearing stratum	Material	Porosity (%)	Specific yield (%)	Hydraulic conductivity (m d^{-1})
Aquifer	Sand, gravel	25–45	25–30	$1–10^4$
	Fractured rock (sandstone, limestone, chalk, igneous and metamorphic)	5–30	1–15	$10^{-1}–10$
	Cavernous limestone and dolomite, basalt lavas	5–40	0.1–5	$10^2–10^4$
Aquitard	Very fine sand, silt, clayey and silty sand	25–60	3–20	$10^{-3}–10^{-2}$
Aquiclude	Clay, shale, mudstone, siltstone, chalk	35–60	1–10	$10^{-8}–10^{-4}$

groundwater in 'exploitable' or 'significant' quantities, though subjective, is nevertheless important as it differentiates aquifers from the much less permeable aquicludes and aquitards. The European Commission definition seems to suggest that a rock need not be water-saturated or water-bearing to be an aquifer. This is may be the more accurate definition, especially in relation to unconfined aquifers in which the water table fluctuates. However, a rock may not be easily called an aquifer without water filling its voids. The UNESCO/WMO description probably provides the simplest and most apt definition. Sands and gravels, fractured limestones, fissured chalk and sandstones, and fractured and weathered volcanics, granites and metamorphic rocks are all examples of geological materials that can form aquifers.

An aquiclude is generally understood to be a saturated but relatively impermeable stratum that does not yield appreciable quantities of groundwater to wells or springs. Clay, mudstone or siltstone or non-fissured shale are examples.

An aquitard is more permeable than an aquiclude but much less permeable than an aquifer. It may be defined as a saturated but poorly permeable stratum that impedes groundwater movement and does not yield water freely to wells or springs. Aquitards can transmit appreciable quantities of groundwater to underlying aquifers, especially under pumping conditions. Examples of aquitards are clayey sands and silty sands.

Types of aquifers

There are three main types of aquifers: unconfined, also known as water table or phreatic (from the Greek *phrear, -atos*, meaning well); confined, also known as artesian; and semi-confined, also known as leaky. In deep sedimentary basins all three types of aquifers can occur together forming a multilayered aquifer system. The idealised section in Figure 2.2 illustrates the occurrence of the three types of aquifers in a multilayered aquifer system.

An unconfined aquifer stores groundwater under atmospheric pressure. Its upper surface is marked by a water table or phreatic surface, which corresponds more or less to the level at which water stands in a well. Groundwater levels are free to rise and fall and their fluctuations reflect changes in the volume of groundwater stored in the pore space. Below the water table is the zone of saturation, and above it to the ground surface is the zone of aeration or unsaturated zone where the pores are mainly filled with air and only partially filled with water. This water is known as vadose water, from the Latin *vadosus* meaning shallow. The unsaturated zone is important as it acts as the conduit to the downward movement of infiltrated water from rainfall or through streambeds to the water table. It has a variable thickness from near zero in areas where the water table is close to the ground surface to more than 100 m in some arid or semi-arid areas. Immediately above the water table is the capillary zone or capillary fringe which extends from the water table to the limit of capillary rise. The capillary rise depends on the grain size of material. In silt it ranges from 100 cm to more than 200 cm, in fine sand to around 43 cm, in medium sand to about 25 cm and in very coarse sand and gravel 2–7 cm (Lohman, 1972).

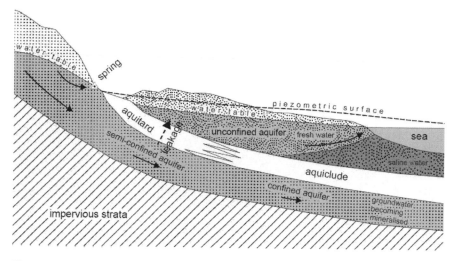

Figure 2.2 Sketch section showing groundwater occurrence

Confined aquifers store groundwater under pressure greater than atmospheric. Their upper surface is bound by an aquiclude, known as the confining layer. Confined aquifers are fully saturated. Fluctuations in groundwater levels reflect changes in pressure, and unlike unconfined aquifers, only minor changes in groundwater storage. Groundwater levels in a confined aquifer are also known as piezometric levels (the prefix *piezo-* is from the Greek *piezein* meaning to press) or potentiometric levels. Artesian aquifers are confined aquifers in which groundwater levels rise above the ground surface. The term is from the French *artesien* referring to Artois, the northernmost province of France, where in about 1750 the first deep flowing wells were drilled and investigated. Figure 2.3 shows an artesian flowing borehole penetrating deep fossil aquifers in the Taoudeni basin of Mauritania.

Semi-confined or leaky aquifers are similar to confined aquifers but their upper confining layer is an aquitard. They are also known as leaky aquifers as groundwater can 'leak' – usually downwards – from the aquitard into the aquifer. Pumping of the aquifer enhances leakage. In deep sedimentary basins where alternations of aquifers and aquitards form multilayered aquifer systems much of the water is derived from leakage.

Transboundary aquifers

Transboundary aquifers, also known as 'internationally shared' aquifers, are aquifers that extend across international boundaries. Stored groundwaters may be fossil or renewable. There are large transboundary aquifers in Africa, the Middle East, South America and Europe. The largest is the Nubian Sandstone aquifer in North Africa which is shared between Chad, Egypt, Sudan and Libya. It stores

Figure 2.3 Artesian flowing borehole in the Taoudeni basin, Mauritania (photo courtesy B. Burnet, 2007)

fossil groundwater and has an area of approximately 2.2 million km^2, a freshwater storage of approximately 170,000 km^3 and exploitable reserves of approximately 15,000 km^3 (Puri, 2001; Foster and Loucks, 2006). Other transboundary aquifers in Africa are the North Western Sahara aquifer shared between Algeria, Libya and Tunisia; the Maastrichtian aquifer of the Senegambian Basin, shared between Mauritania, Senegal and Gambia; and the Kalahari-Karoo multilayered aquifer system which is shared by Namibia, Botswana and South Africa. In the Middle East the main transboundary aquifer is the Disi–Saq sandstone aquifer, shared between Jordan and Saudi Arabia. The Guarani transboundary aquifer system is the largest in South America. It is shared by Argentina, Brazil, Paraguay and Uruguay and forms a huge source of fresh water. It is rechargeable and has an area of 1.2 million km^2 and a storage volume of 40,000 km^3. There are three main transboundary aquifers in Europe, the Vechte aquifer in Western Europe and the Slovak Karst-Aggtelek and Praděd aquifers in Central Europe.

Hydraulic properties of water-bearing strata

The main hydraulic properties that characterise water-bearing strata are porosity, storage coefficient (storativity), specific yield and hydraulic conductivity. Typical general values from various sources are presented in Table 2.1. Hydraulic properties may be determined in the laboratory and in the field. In groundwater evaluations, field values, especially those from pumping tests on wells, are preferable as they generally apply to a larger area of the aquifer. A pumping test (or aquifer test) involves the pumping of a well usually at a constant rate of discharge, for a period of time while observing groundwater levels in observation wells in the surrounding area. The analysis of the time-drawdown or distance-drawdown data provides estimates of hydraulic parameters and also an understanding of aquifer behaviour. Kruseman and de Ridder (1992) provide an excellent description of pumping test analysis.

Porosity and effective porosity

Porosity (Φ) provides a measure of the void or pore space of rocks and soils. It is expressed as the ratio of the volume of voids (V_v) to the total volume of the rock (V_r):

$$\Phi = \frac{V_v}{V_r} \text{ (expressed as a fraction or per cent)}$$

The interconnected void space available for groundwater flow is known as effective porosity and is more important than total porosity. Clays, silts and chalk have a very high total porosity but very low effective porosity, and therefore flow of water is slow and in very small quantities. Porosity that results from spaces (pores) between grains is known as primary or intergranular porosity. Rocks in which voids are due to fractures, joints or other discontinuities are said to possess secondary porosity. In these rocks the solid rock is practically impermeable and groundwater flows through the passages formed by the discontinuities. But there are rocks – chalk is a good example – which have dual porosity. In these, the solid rock (or matrix) is porous and groundwater is stored in both the porous rock and in fractures.

Storage coefficient (storativity) and specific yield

Storage coefficient (S), or in unconfined aquifers specific yield (S_y), is a dimensionless quantity defined as the volume of water that is released from or taken into storage per unit surface area for a unit change in hydraulic head that is normal to that surface. In confined aquifers, the release of water from storage when the hydraulic head declines is the result of the compaction of the aquifer and the expansion of the water. Thus, the values of storage coefficient are small, in the order of 10^{-3} to 10^{-5}. In unconfined aquifers, the release of water from storage when the hydraulic head declines corresponds to the drainage of the interconnected pore space under gravity and is therefore similar to effective porosity. As can be seen from Table 2.1, specific yield ranges from 1 per cent (0.01) to 30 per cent (0.3), and is much greater than the storage coefficient of confined aquifers.

Permeability and hydraulic conductivity

Permeability and hydraulic conductivity are physical properties related to the ability of rocks to transmit a fluid. But whereas permeability, also known as intrinsic permeability (k), is a property of only the rock or soil, hydraulic conductivity (K)

includes the properties of the fluid. Permeability has units of area (L^2) and is used primarily in the petroleum industry. Hydraulic conductivity has units of length time^{-1} (LT^{-1}) which are the same as those of velocity usually expressed as md^{-1}. In groundwater, hydraulic conductivity is defined as

> the volume of water at the prevailing kinematic viscosity that will move through a porous medium under a unit hydraulic gradient through a unit area measured at right angles to the direction of flow.

Values of hydraulic conductivity for different geological materials are shown in Table 2.1. In unconsolidated granular media, hydraulic conductivity varies with particle size and degree of sorting; clays and silts have low values and the coarser grained sands and gravels higher values. In consolidated and cemented rocks hydraulic conductivity is dependent on the degree of fracturing.

Transmissivity

Transmissivity (T) is widely used in groundwater evaluations and is the property that is usually derived from the analysis of pumping tests. It is defined as the rate at which groundwater is transmitted through a unit width of aquifer under a unit hydraulic gradient. It has the units of L^2T^{-1} and is usually expressed as m^{2}d^{-1}. Very productive aquifers have values of transmissivity of several thousand m^{2}d^{-1}, average aquifers about hundred to a thousand and poor aquifers around a hundred or less. Transmissivity is obtained by multiplying hydraulic conductivity by the aquifer saturated thickness (b):

$$T = kb$$

The relationship suggests that transmissivity varies linearly with aquifer thickness, i.e. the thicker the saturated aquifer, the greater the transmissivity. In reality the relationship does not always hold. In thick aquifers permeability tends to decrease with depth due to compaction and in fractured or weathered aquifers only the upper part is generally permeable.

Hydraulic head and groundwater movement

Hydraulic head is an important concept in hydrogeology and has already been referred to in the discussion on hydraulic parameters. In slow moving groundwater, it may be represented by the following equation:

$$H = z + \frac{P}{\gamma g}$$

where H is hydraulic head (L)
   z is the groundwater elevation in relation to a datum (L)
   P is the pressure exerted by the column of water (ML^{-2})

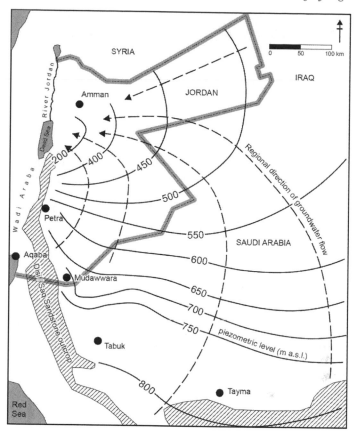

Figure 2.4 Idealised piezometry and groundwater flow. Example from the Disi-Saq transboundary sandstone aquifer (adapted from HSI, 1990)

γ is the specific of water (ML^{-3})

g is the acceleration due to gravity (LT^{-2})

In practical terms, hydraulic head is the height that groundwater will rise in a well in relation to a datum (normally, sea level). For example, if the water level in a well is 50 m below the ground surface and the ground surface is 150 m above sea level, the hydraulic head is 100 m above sea level; and if the water level is 50 m above the ground surface, the hydraulic head is 200 m above sea level. In confined/semi-confined aquifers the hydraulic head is equivalent to the piezometric level and in unconfined aquifers to the water table.

Groundwater in aquifers is generally in motion. It moves from areas of high hydraulic head to areas of low hydraulic head. Piezometric maps or groundwater level contour maps show the distribution of hydraulic heads and indicate the direction of groundwater movement. An example from the transboundary Disi-Saq sandstone aquifer is shown in Figure 2.4 (Hydrogeological Services International – HSI, 1990). The map shows groundwater to move from the outcrop areas over distances of several hundred kilometres to emerge along the low lying Dead Sea. As recharge in the last 10,000 years has been negligible, groundwater movement is driven by relic hydraulic heads of long ago.

Percolation and seepage

Percolation is not a term that is nowadays often used in hydrogeology. Meinzer (1968) described it

> as the movement of groundwater through the interstices of rock or soil under hydrostatic pressure, but not including the movement of water through large openings such as caves or the flow in subterranean streams such as found in karstic limestone areas.

Percolation is also used to describe the movement of water past the soil zone and through the unsaturated zone to the water table. Seepage refers to the slow and diffuse movement of groundwater emerging at the ground surface. In seepage areas water may pond and evaporate, or flow, depending on the magnitude of the seepage, climate, and the topography. The term is well established in the engineering literature in connection with groundwater movement in excavations, canals and dams.

Darcy's law

Darcy's law states that the flow rate through porous media is proportional to the hydraulic head loss and inversely proportional to the length of the flow path. In the form of an equation it is expressed as:

$$\frac{Q}{A} = \frac{K(H-h)}{l}$$

where Q is the flow rate (L^3T^{-1})
K is the proportionality constant, the same as hydraulic conductivity
A is the cross-sectional area (L^2) through which flow takes place
$H–h$ is the difference in hydraulic head between two points along the direction of groundwater flow (L)
l is the length of the flow path (L)

The term $(H–h)/l$ is known as the hydraulic gradient (i) and is a dimensionless quantity. Hydraulic gradients in aquifers can be estimated from piezometric (groundwater level) contour maps. The term Q/A has the same dimensions LT^{-1} as velocity (v). Thus

$$v = Ki$$

This is known as the 'Darcy velocity' to distinguish it from the true velocity (u) that takes place through the connected pore space or effective porosity (φ_e)

$$u = \frac{v}{\varphi_e}$$

Clearly the velocity through the pores is always larger than the Darcy velocity. For example the Darcy velocity of groundwater moving under a hydraulic gradient of 0.005 in sand having a hydraulic conductivity of 100 md^{-1} is 0.5 md^{-1} whereas the average velocity through the pores assuming an effective porosity of 20 per cent is 2.5 md^{-1}, i.e. five times greater.

Darcy's law is valid for low velocities where flow is laminar. Most groundwater flow is slow and within the laminar flow regime. As the flow velocity increases turbulence sets in and Darcy's law is no longer valid. This is the case in fast-flowing groundwater in fractures (fissure flow), caverns and large openings and also close to pumping wells where hydraulic gradients become very steep. Darcy's law may also not be valid in very fine materials such as clays.

Darcy's law may be used to determine the horizontal groundwater flow in aquifers. A convenient equation is

$$Q = Tiw$$

Q is the rate of groundwater flow (L^{3}T^{-1}) and w(L) is the width of aquifer through which flow takes place. The hydraulic gradient (i) is obtained from groundwater level contour or piezometric maps. The equation provides a quick method for estimating groundwater flow providing influences due to changes in groundwater storage are not significant. Useful applications are assessing groundwater recharge and in transboundary aquifers the flow across national boundaries.

Groundwater flow velocities

Groundwater velocities vary from a few metres per year to a few metres per day. In fractured basalts and karstic limestones groundwater can travel very fast, more than a few hundred metres per day. In large fossil groundwater basins velocities can be very low and it may take thousands of years for groundwater to reach its discharge point. Also in deep confined aquifers, groundwater moves extremely slowly and may be almost stagnant.

Groundwater flow around pumped wells

Pumping of groundwater from an aquifer causes groundwater levels in the surrounding area to be lowered. The difference between the original groundwater level and the new (pumped) groundwater level is known as 'drawdown'. In three dimensions the drawdown-distance curve has the shape of a cone, and therefore the area of influence of the pumped well is known as the 'cone of depression'. Figure 2.5 shows the cone of depression in a confined and semi-confined (leaky) aquifer and Figure 2.6 in an unconfined aquifer.

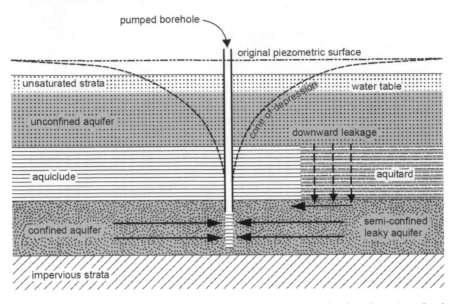

Figure 2.5 Cone of depression around a pumped well in a confined and semi-confined (leaky) aquifer

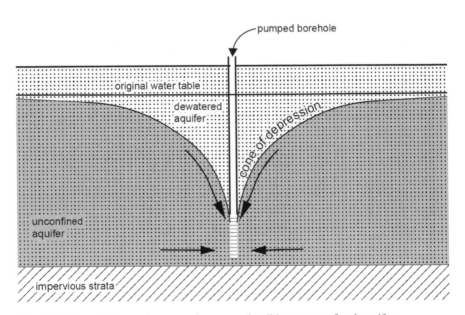

Figure 2.6 Cone of depression around a pumped well in an unconfined aquifer

As the cone of depression expands it may intercept surface water bodies, such as rivers, lakes or the sea. Induced water from these sources or from downward leakage from aquitards can reduce drawdowns or stop the cone of depression from expanding altogether. When drawdowns no longer change with the time of pumping, groundwater levels reach a state of equilibrium and flow conditions are said to have reached a steady state. Without an external source of water, groundwater is taken from aquifer storage and groundwater levels continue to fall, never reaching equilibrium. Under these conditions, flow is said to be unsteady or transient. In practice, in extensive aquifers, as the cone of depression expands, drawdowns become smaller and smaller and a near equilibrium state or quasi-steady state is reached.

Equations to determine the drawdown in confined aquifers for steady flow conditions were developed by Thiem (1887) and for unsteady state flow conditions by Theis (1935). Cooper and Jacob (1946) simplified the Theis equation, Hantush (1956) developed a solution for semi-confined leaky aquifers and Boulton (1963) for unconfined aquifers. These and many other equations and graphical techniques arising from them have been used to determine the hydraulic properties of aquifers and aquitards, and can be found in most textbooks (Kruseman and de Ridder, 1992; Todd and Mays, 2005).

Groundwater discharges

In shallow water tables of less than about one metre in depth, groundwater is lost directly to evaporation. Groundwater from greater depths can be taken up by the roots of deep penetrating plants or phreatophytes (from the Greek *phrear -tos* meaning well or spring and *phyton* meaning plant). In river valleys or depressions where the water table intercepts the ground, surface groundwater emerges as seepages or springs. The headwaters of some rivers are fed from springs. A large proportion of the baseflow of rivers is from groundwater discharges from springs and seepages into the river channel. Springs and seepages also form the source of water to many wetlands and oases. Many of the North African *sebkhas* (salt pans) have formed as result of evaporating groundwater discharges.

Springs

Springs are concentrated groundwater flows discharging at the ground surface but also at the sea bottom (submarine springs). Their flow may vary from a few litres per second to many thousands, depending on catchment area. The larger springs are found in fractured volcanic rock and karstic limestones. Groundwater may travel to springs from distant parts of the aquifer, which may lie well beyond the surface water catchment boundaries. Flows fluctuate seasonally in response to rainfall. Springs from karst or fractured aquifers respond quickly, often within a few hours or a few days after rainfall has occurred. Springs originating from confined aquifers which have distant recharge areas take much longer. Many springs are perennial with some dating back to ancient times. Many old cities

(Damascus, Jerusalem, Athens and Rome) used springs as their source of drinking water. There are many hydrogeological settings that give rise to springs: where the water table intercepts the ground surface as mentioned above; via faults or deep fractures tapping confined aquifers; or via conduits in fractured rock or karstic limestone. Submarine springs as sources of fresh water have been known since ancient times in the coastal areas of the Mediterranean and are still flowing today off the coast of Syria, Lebanon and Greece.

There are submarine springs from limestones in the eastern coast of the United States (Florida), the coast of the Yucatan province in Mexico and from fractured volcanics in Hawaii. Submarine springs emerge from aquifers that extend offshore beneath the sea bottom. In confined aquifers groundwater emerges at some distance from the shore either through conduits in the overlying aquiclude/ aquitard or at the edge of the 'outcropping' aquifer. Submarine springs from unconfined aquifers are usually closer to the shore and more brackish. Global estimates of the flow of submarine springs vary widely from 0.2 to 10 per cent of total river flow (Taniguchi *et al.*, 2002). Thermal springs consist mainly of meteoric water which has infiltrated deep into rocks to become heated and rise to the surface. Geysers are thermal springs that flow intermittently. Thermal springs are found in all parts of the world, the more famous being those in Iceland, New Zealand, the Kamchatka peninsula in Russia and Yellowstone Park in the United States.

River–aquifer interactions

The various types of river–aquifer interactions are illustrated in Figure 2.7. In fissured and karstic limestone aquifers disappearing and reappearing rivers are well known. River water enters fractures to reappear at a lower level downstream. Rivers flowing over permeable strata may lose or gain water depending on the position of the river water level relative to the groundwater level. Where the river water level is higher than the groundwater level, the river loses water to the aquifer; and where it is lower, the river gains groundwater from the aquifer. Also, where the river and aquifer are not in direct hydraulic continuity, i.e. 'disconnected' or 'detached', water is lost from the river to the aquifer. River water travels through the unsaturated zone to reach the water table.

One of the impacts of excessive groundwater abstraction is a reduction in river flow, and in extreme cases the drying up of the upper reaches of tributary streams, due to a general lowering of groundwater levels. In the UK good examples are the Chalk and the Sherwood sandstone aquifers (Owen, 1991; Rushton and Tomlinson, 1995). River flow losses and gains are estimated by measuring the flow at different locations and establishing depletion or accretion profiles. River–aquifer interactions can be complex as they may be influenced by riverbed permeability, which may be variable, aquifer characteristics and water level changes. In recent years, modelling techniques have been used. Most have been based on finite difference models, such as MODFLOW, or simpler spreadsheet formulations using the Theis formula or variations of it (Younger, 1995; Parkin *et al.*, 2002).

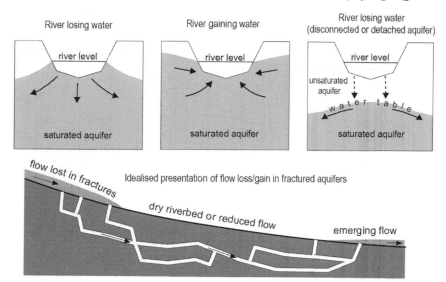

Figure 2.7 River–aquifer interactions

Sophocleous *et al.* (1995) applied simplified stream–aquifer depletion models in water rights administration.

Groundwater recharge

Natural recharge

Groundwater recharge is the process by which aquifers are replenished. Under natural conditions aquifers are recharged by the infiltration of rainfall or melting snow (direct recharge) or by infiltration of surface water through the beds of rivers, streams or lakes (indirect recharge).

The infiltrated water travels through the unsaturated zone. It may take from a few days to many months or even years to reach the water table. In karstic regions or where fractured aquifers are exposed at the ground surface, infiltration is rapid, with most of the rainfall reaching the water table directly without having to travel through the unsaturated zone. Infiltration of surface water through streambeds is the dominant recharge mechanism in arid areas where evaporation is very high (2,000–4,000 mm a^{-1}) and rainfall is low, less than 100 mm per year. Most recharge occurs through river beds where runoff concentrates following intense rainfall events. The amounts depend on the permeability of the subsurface materials, the slope of the river bed, flow velocity and the depth of the underlying water table.

There are a number of methods used to estimate recharge. Although all appear straightforward there is considerable difficulty in evaluating individual parameters, so estimates are always at best approximate. A comprehensive description of methods has been presented by Lerner *et al.* (1990) and more recently by Healey

(2010). As mentioned above, the Darcy equation provides a relatively easy method in large aquifers when flow is at steady state. In closed basins, gains or losses in groundwater storage may be calculated from changes in groundwater level (Δh) multiplied by aquifer storativity (S). More exotic techniques involve the use of chloride and isotopes (Edmunds and Walton, 1980). Recharge is often calculated by means of the water balance (or water budget) method, which may be stated as

$$R = P - \text{Eta} - \text{Ro} \pm \Delta S$$

where R is recharge
         P is precipitation (rainfall and snow)
         Eta is actual evapotranspiration
         Ro is river flow excluding baseflow
         ΔS is change in groundwater storage

All parameters are in units of L, usually expressed in mm.

P and ρ are usually measured parameters, though Ro is derived by hydrograph analysis of the measured total river flow (R_y); Eta is a function of many factors (temperature, wind speed, radiation, humidity, soil moisture, crop type etc). Because it is difficult to measure it, mathematical models have been developed which relate potential evaporation (Penman, 1948) with soil moisture (Monteith, 1965; Grindley, 1969). A standardised method for determining evapotranspiration known as the FAO Penman–Monteith method was developed by Allen *et al.* (1998).

The term P–Eta is known as 'hydrologically effective rainfall' or simply 'effective rainfall' and provides a measure of the amount of rainfall that is available for infiltration into aquifers and for river flow. As evapotranspiration is generally greater in the summer and rainfall lower, recharge often takes place mainly during the winter months. There are two components to river flow: a surface water component, which represents runoff (Ro in the above equation) and baseflow, which is groundwater from springs or seepages discharging into the river channel. For catchments dominated by permeable rock there is little surface water runoff. In some of the 'dry valleys' of the English Chalk more than 90 per cent of river flow may be derived from groundwater. In low permeability clay catchments most of the river flow is from surface water runoff with very little contribution from groundwater.

Descriptions and examples of estimates of groundwater recharge in different hydrogeological provinces (alluvial fans and river beds, sand and sandstone, limestone and dolostone, chalk, volcanic and plutonic crystalline) have been given by Lerner *et al.* (1990).

Global natural recharge estimates

Global groundwater recharge for the period 1961–1990 was estimated by Döll and Fiedler (2008) to be 12,666 km^3 a^{-1}, which corresponds to approximately 95 mm a^{-1} or approximately 12 per cent of rainfall. South America receives the highest recharge approximately 230 mm a^{-1} and Australia the lowest at 35 mm a^{-1}.

Africa and North and Central America receive similar recharge, approximately 70 mm a^{-1}, Asia 100 mm a^{-1} and Europe 50 mm a^{-1}. As would be expected, the distribution of recharge in each continent varies considerably from less than few millimetres in the desert areas (Sahara in North Africa, Gobi in Central Asia, Kalahari in Southern Africa and the deserts of Central and Western Australia and western USA) to more than 1,000 millimetres in the tropical regions of the Amazon in South America, and of Central Africa and Southeast Asia. In dry regions (Arabian Peninsula, North Africa, western USA, Chile, Western Australia, southern Spain) recharge is small, generally less than 20 mm a^{-1}. Scanlon *et al.* (2006) provided a global synthesis of groundwater recharge in arid and semi-arid regions. They estimated average recharge rates over large areas (40,000–374,000 km^2) in the range of 0.2 to 35 mm a^{-1}, representing 0.1–5 per cent of long-term average annual precipitation.

Irrigation returns and urban leakage

Recharge also occurs due to human activities. In agricultural areas from irrigation returns (canals and fields) and in urban areas from septic tanks and latrines, leaking sewers and leaking water mains (Lerner *et al.*, 1990). In flood and furrow irrigation, as practised in paddy-field rice in the Far East and, mainly in the past, for the growing of vegetables, and fruits in the Middle East and the Mediterranean countries, a substantial part of the applied water (up to about 40 per cent) can infiltrate to the subsurface, and given favourable subsurface conditions, reach the water table. In hot climates, evaporation of the rising water table can cause soils to become saline and unproductive. In unsewered towns or cities, infiltration from septic tanks and latrines can form a significant proportion (up to 60–70 per cent) of the recharge to the groundwater (Hydrogeological Services International, 1998). Leakage from sewers and water supply pipes can also be substantial. Many of the older pressurised water distribution systems lose 20 to 35 per cent of their water. Rising groundwater levels and groundwater flooding in many cities of the Middle East have been the result of recharge of shallow unconfined aquifers from leaking water and sewerage pipes.

Artificial recharge

Artificial recharge involves the introduction of water into aquifers by artificial means. It can be achieved either indirectly by allowing water to infiltrate through basins, ditches, dykes in river beds, dams, and various other water-spreading methods, or directly by injecting water into boreholes. A variant of direct artificial recharge is aquifer storage recovery (ASR) in which fresh water is injected through boreholes into slow-moving brackish or saline groundwater in a confined aquifer (Pyne, 2005). The fresh water forms a 'bubble' on top of the brackish/saline groundwater, which is extracted for use when required. Recharge dams that temporarily impound flood runoff and release it slowly to allow it to infiltrate into the underlying aquifers have been used in the Middle East. In the Far East, the United States and also the Middle East, dykes and surface or subsurface check dams in river beds have been used to increase the resident times of flows and, thus, increase the quantities of infiltration into the underlying sand aquifers. Basin recharge has been extensively used in the USA, especially in California and Long Island, New York, and in Israel (Todd and Mays, 2005).

Effects of climate change on groundwater recharge

It should not be surprising that climate changes, whether due to human activities or natural causes, have effects on both surface water and groundwater. For example, the large groundwater reserves in the fossil aquifers of arid North Africa and the Middle East bear witness to a more humid climate in the not so distant past, and rates of recharge far greater than the meagre quantities of today. The two hydrological parameters of importance are temperature and precipitation. The first affects evapotranspiration, the higher the temperature, the greater the evapotranspiration, and the second, the input to the recharge process. The combined effects of increased evapotranspiration and decreased precipitation result in a reduction in potential recharge. However, natural scenarios are not usually so simple. Thus, although global warming may generally lead to increases in air temperature, precipitation may increase or decrease, be greater in the summer and less in the winter or vary in intensity. Each of these changes may have a different effect on groundwater recharge. Taken together with other factors, such as the nature of soils and the unsaturated zone, types of vegetation and land use, the task of predicting changes to recharge patterns in space and time becomes very difficult. Moreover, global climate models (GCM), although sophisticated, operate on a coarse scale, and are not easily applicable to smaller areas. There are downscaling methods to do this, which are generally of two types: (1) dynamic climate modelling and (2) empirical statistical downscaling (Green et al., 2011). At present, most downscaling is empirical. It uses statistical techniques to obtain relationships between observed regional or local climate variables and large scale climate variables. Downscaled daily temperatures compare well with the observed data, but not daily rainfall, seasonal amounts of rainfall or lengths of wet periods (Green et al., 2011).

Until very recently much of the research work on the impact of climate change has focused on surface water, which is visible and less complicated than groundwater. But there has been a surge in interest in the last eight years, notably during 2006 to 2009, when the number of peer-reviewed journal papers has more than doubled (Green et al., 2011). In general, most studies suggest that recharge rates may decrease as a result of climate change but there are numerous others which indicate that in some areas it may actually increase or not change at all. Also, other factors such as deforestation or abstraction may in some cases be more significant than climate change. A global synthesis of 20 case studies on the effects of climate change on groundwater has been presented by Treidel et al. (2012). Some of the main conclusions are:

1   in some tropical wet regions and also in semi-arid areas, such as the High Plains aquifer in the USA, where infiltration is from runoff accumulating

in ponds, gullies and depressions, recharge rates may increase due to the projected increase rainfall intensity;

2   in some of the tropical-dry regions the projected decreases in rainfall and increases in air temperature may result in decreases in groundwater recharge rates – in Mali, West Africa, recharge may decrease by 8–11 per cent depending on soil properties;

3   in low-lying countries, such as the Netherlands and in the carbonate Pacific islands, the major concern is rising sea levels, particularly in relation to low-lying atolls, which may contaminate groundwater, especially during droughts;

4   in countries with continental climates, such as China, an increase in rainfall variability may induce droughts, which will affect recharge, while at the same time tending to increase groundwater abstraction.

In the same publication, Hiscock et al. (2012) in their evaluation of future climate change impacts on European groundwater resources, comment that the predictions from most GCMs indicate that winter groundwater recharge in northern Europe by the end of the century will increase, but become limited during the longer summers. In southern Europe, especially southern Spain, groundwater recharge is expected to decrease overall.

Groundwater quality and pollution

Groundwater quality has now become as important as quantity. This is partly because of threats to ecosystems and human health but also because degradation in water quality affects usable reserves and therefore quantity. Standards relating to the use of water for drinking are relatively new. European and international drinking water standards were published by WHO in 1958, 1961 and 1971. These were superseded in 1984 with the issue of guidelines for drinking water quality, which set guideline values for a large number of constituents or parameters, including organic, inorganic and microbial constituents of health significance. Three further editions followed, the latest, is the fourth edition of 2011, which includes all the amendments and additions of the previous editions and a section on radiological parameters (WHO, 2011). In 1980 the European Commission (EC) issued Directive 80/778/EEC 'relating to the quality of water intended for human consumption', which listed standards for 66 parameters for which guide (G) values and maximum admissible concentration (MAC) values were stipulated (European Commission, 1980). In England and Wales, the EC Directive was implemented by the enactment of the Water Supply (Water Quality) Regulations 1989, and in Scotland through amendments of the Water (Scotland) Act 1980 and the Water Supply (Water Quality (Scotland) Regulations, 1990. The EC Drinking Water Directive has been progressively incorporated by all the European Union member states. In the USA, water quality standards arose from the Safe Drinking Water Act 1974, which directed the US Environmental Protection Agency (US EPA) to establish minimum drinking water standards (Gray, 1994). Standards were issued by US EPA in 1993 and National Drinking Water Regulations in 2010. Standards and guidelines have been undergoing various revisions and amendments to reflect both progress in analytical techniques and the knowledge of potential risks to health. WHO guidelines (WHO, 1984 and subsequent editions) have been used as the basis for national standards in many countries. Standards in relation to other uses (industrial, irrigation, livestock watering) have been described in Lloyd and Heathcote (1985).

Major ions and other constituents

Groundwater contains a large number of chemical constituents in solution. Most of them are natural but an increasing number are man-made. Traditionally, dissolved constituents have been divided in two categories: major ions (ions are electrically charged atoms or molecules), which make up the bulk of dissolved constituents, and minor constituents or trace elements, which occur in very small concentrations. Major ions include the cations (positively charged atoms) calcium, magnesium, sodium and potassium, and the anions (negatively charged atoms or radicals) chloride, sulphate, carbonate and bicarbonate and, depending on concentration, sometimes nitrate. Silicon can also be an important constituent, generally as dissolved silicon dioxide (silica). There are several minor constituents but the more important include fluoride, some of the heavy metals (iron, manganese, lead, nickel, arsenic, chromium, cadmium) and lighter metals, such as aluminium, boron, phosphorus and strontium. In the past, routine chemical analyses of groundwater for potable supply examined primarily the major ions, nitrate and fluoride, microbiological parameters (total coliforms, *E coli*) and periodically heavy metals. In recent years the list has been extended to include the very large number of manmade organic compounds which have entered the groundwater in many aquifers as a result of industrial and agricultural activities, notably industrial solvents (chlorinated volatile organic compounds) and pesticides. Although these compounds generally occur in very low concentrations they can be highly toxic.

Total dissolved solids (TDS)

Total dissolved solids (TDS) represents the total of mineral constituents, ionic and non-ionic, dissolved in water. It is normally the solid residue after water has been evaporated to dryness at 180°C. The TDS of fresh groundwater is generally taken to be less than 1,000 mg l^{-1}, brackish groundwater 1,000–10,000 mg l^{-1} and saline (salty) groundwater 10,000–100,000 mg l^{-1} (Gorrell, 1958). Seawater has a TDS of about 35,000 mg l^{-1} and hypersaline waters or brines more than 100,000 mg l^{-1} and as high as 300,000 mg l^{-1}. Brines are found in evaporating enclosed seas, such as the Dead Sea, and in shallow aquifers in *sebkha* areas and arid coastal regions or at depth in oil reservoirs, near salt domes.

Physicochemical parameters

The main physicochemical parameters that usually characterise groundwater are electrical conductivity (EC), hydrogen ion activity (pH) and the redox potential

(Eh). Electrical conductivity is a measure of how easily an electric current passes through a liquid. It is usually measured in microsiemens per centimetre (μS cm^{-1}). Very pure water has an EC value approaching 0.05 μS cm^{-1} and distilled water at least 1.0 μS cm^{-1} (Hem, 1985). Groundwater has a range of EC values varying from as low as 50 μS cm^{-1} to many thousand. Fresh water has an EC of less than about 1,500 μS cm^{-1}, sea water 50,000 μS cm^{-1} and brine more than 140,000 μS cm^{-1}. Because most constituents in groundwater occur in ionic form, EC is related to the concentration of TDS. For most groundwaters the relation between EC and TDS is more or less linear and is often used to obtain an approximate value of TDS. Stated as an equation

$$TDS = cEC$$

where TDS is expressed in mg l^{-1} and EC as μS cm^{-1} at 25°C . The value of the constant c is on average 0.7, but may vary for different groundwaters between 0.55 and 0.8 (Lloyd and Heathcote, 1985).

Neutral water contains 10^{-7} moles per litre of hydrogen (H$^+$) and 10^{-7} moles per litre of hydroxyl (OH^{-1}) ions. The negative logarithm to the base of 10 of the concentration of hydrogen ions is known as the pH. Thus, a solution is neutral at a pH of 7, acidic at a pH less than 7 and alkaline at a pH greater than 7. Most groundwaters have a pH in the range of 5 to 8.

Redox potential (Eh) indicates whether water is oxidising or reducing. It is expressed in units of millivolts (mV). Generally a large positive value of a few hundred millivolts indicates an oxidising environment in contact with the atmosphere and a negative value reducing conditions isolated from the atmosphere. In natural environments the value of Eh is strongly influenced by pH, as pH increases Eh decreases (Garrels and Christ, 1965; Appelo and Postma, 2005).

Rainwater and groundwater

As already mentioned, rainfall is the main source of water entering both rivers and aquifers. Rainwater is a very dilute slightly acidic solution with a TDS of generally 4–10 mg l^{-1} and a pH of 5–6. Its main constituents are sodium, chloride and sulphate, their concentration generally decreasing away from the seacoast. The sodium/chloride ratio of rainwater resembles that of the ocean. Nitrate concentration in rainwater is very low, less than 1 mg l^{-1}. It forms by the oxidation of atmospheric nitrogen in electric storms and of nitrogen oxides emitted in the exhaust fumes of vehicle engines. The dissolution of industrial gases, mainly sulphur dioxide and nitrogen oxides, by falling rain in England, Scandinavia and the northeastern USA gave rise to acid rain in the 1950s and 60s with devastating effects on some woodland and forest areas (Cowling, 1982; Hem, 1985).

Newly infiltrated rainfall in shallow sand aquifers has a composition that is close to rainfall and a TDS of about 50 mg l^{-1}. Groundwaters with long residence times are generally more mineralised, with total dissolved solids of several thousand mg l^{-1}. Exceptions are groundwaters stored in mainly quartzose sandstones with little soluble matter. Groundwaters in limestone and dolomite aquifers or in calcareous sandstones are hard and rich in calcium, magnesium and bicarbonate; groundwaters in fractured volcanic and basement rocks or quartzites are soft with higher silica and less calcium. The presence of soluble salts such as sodium chloride, gypsum or anhydrite results in brackish groundwaters rich in sodium, chloride and sulphate. In multilayered aquifer systems, groundwaters in different layers can have different chemistries. Where layers are hydraulically connected there may be mixing of different types of groundwaters. Mixing also occurs during the upconing of underlying saline water in pumped wells and due to seawater intrusion of coastal aquifers. Mixing is often accompanied by chemical reactions so that the end product is not always a simple mixture of components as might be the case with the conservative chloride ion. In complex systems computer programs have been developed to determine mixing effects (Parkhurst *et al.*, 1980; Plummer *et al.*, 1994).

Chemical changes along the direction of groundwater flow

In general, the chemical character of groundwater changes as it moves from unconfined conditions in the recharge area to confined conditions at depth, where flow is very sluggish. The main change concerns oxygenated calcium bicarbonate groundwater progressively giving way to sodium rich anoxic groundwater (Chebotarev, 1955). This is brought about by two main hydrochemical processes: cation exchange and redox reactions. In cation exchange calcium in groundwater is exchanged for the sodium which may be present in dispersed clays within the aquifer or in the overlying aquiclude/aquitard. Redox reactions involve the reduction in the presence of organic matter initially of dissolved oxygen and then of nitrate, manganese and iron oxides, and sulphate. The result is a groundwater that is almost entirely devoid of dissolved oxygen, has a strongly negative redox potential (Eh) and only a trace of nitrate, most of it having been converted to ammonia. Sulphide, having replaced sulphate becomes dominant, which gives groundwater discharging from wells the characteristic rotten egg odour of escaping hydrogen sulphide gas, and iron and manganese go into solution in the form of ferrous iron (Fe^{++}) and manganese (Mn^{++}).

Radionuclides in groundwater

Radionuclides are hazardous to health due to the emission of harmful radiation. The more important naturally occurring radionuclides in groundwater are uranium (U^{238}), its daughter elements radium (Ra^{226}) and radon (Rn^{222}), which is a gas. All three emit α particles and radon also emits β particles. Uranium has the longest half life, 4,500 million years, and radon the shortest, 3.83 days. Radium has a half life of 1,620 years. Uranium is found in most groundwaters in concentrations between 0.1 and 10 micrograms per litre (μgl^{-1}). In groundwater moving through rocks rich in uranium concentrations can be greater than 1.0 μgl^{-1} (Hem, 1985). Radium (Ra^{226}) is extremely toxic, but due to its general immobility is not abundant in groundwater. Concentrations are mostly below 1.0 picoCurie per litre (pCi l⁻¹) (0.037 Becquerel per litre, Bql⁻¹), but generally higher in deep groundwaters or in uranium or phosphate mining areas. In the Nubian Sandstone in the Negev, Israel and in the Disi Sandstone, Southern Jordan, mean values of Radium 226 were found to range between approximately 0.26 and 0.68 Bql⁻¹ and Radium 228 between 0.42 and 1.99 Bql⁻¹ (Vengosh *et al.*, 2009). Radon forms from the disintegration of Ra^{226} in the aquifer rock, and is generally predominant in granitic terrains. Radon dissolves easily in groundwater where it stays until it is exposed to the atmosphere or it decays. In enclosed spaces where it cannot be lost to the atmosphere, it may constitute a serious health hazard and has led to concern in many countries as a cause of lung cancer. In a national survey of groundwaters in the USA the average level of radon was 900 pCi l⁻¹ and the median 300 pCi l⁻¹, but much higher concentrations in excess of 100,000 pCi l⁻¹ were also found in some states (Dupuy *et al.*, 1992; Helms and Rydell, 1992). Surveys in member states of the European Union have shown that radon concentrations in groundwater vary from 1 to 50 Bq l⁻¹ (1 Bq l⁻¹= 27.027 pCi l⁻¹) for rock aquifers in sedimentary rocks, 10 to 300 Bq l⁻¹ for wells dug in soil, and 100 Bq l⁻¹ to 50,000 Bq l⁻¹ in crystalline rocks (European Commission, 2001). In Sweden, radon concentrations in granite groundwaters range from 4 to 400 Bq l⁻¹, in uranium rich granites from 300 to 4,000 Bq l⁻¹ and a maximum of 89,000 Bq l⁻¹ (Skeppström and Olofsson, 2007). The maximum contaminant levels (MCL) for uranium and radium (226 and 228 combined) in drinking water in the US are 30 μg l⁻¹ and 5 pCi l⁻¹ (0.185 Bq l⁻¹) respectively (US EPA, 2009). The World Health Organization guidance level for uranium is 10 Bq l⁻¹ and for Ra^{226} 1 Bq l⁻¹. For Ra^{228}, it is much lower, 0.1 Bq l⁻¹. Because much of the radon in solution is lost to the atmosphere, the guidance level given by WHO is 100 Bq m⁻³ for indoor air in dwellings (WHO, 2011).

Environmental isotopes

Environmental isotopes have been used in hydrogeology to investigate problems especially in relation to the age and origin of groundwater. The most commonly used isotopes are the radionuclides tritium (H^3) and carbon-14 (C^{14}) and the stable isotopes oxygen-18 (O^{18}), deuterium (H^2 or D) and carbon-13 (C^{13}). The use of environmental isotopes in groundwater has been comprehensively described in Kazemi *et al.* (2006).

Tritium is the heaviest isotope of hydrogen and has a half life of 12.43 years. There are two sources to tritium: the first is from the interaction of cosmic rays in the upper atmosphere with nitrogen, which produces 5 to 20 tritium units (TU) in rainfall; the second is mainly from the detonation after 1952 of thermonuclear devices in the atmosphere, which led to substantial increases of several hundred to a few thousand TU in rainfall. The application of tritium to groundwater dating and recharge estimates has been limited, partly because of its short half life and partly because with the moratorium on atmospheric detonations since the early 1960s, atmospheric tritium levels have steadily decreased. It has, nevertheless, been useful for a time in indicating the presence of recent 'post-bomb' water in aquifers, mainly until the end of the 1960s, when tritium concentrations remained fairly high.

Carbon has three naturally occurring isotopes: the radioactive carbon-14, the stable isotope carbon-13, and the most abundant of the three (98.89 per cent) carbon-12. Carbon-14 has a half life of 5,370 years. It is produced in the same way as tritium by the reaction of cosmic rays with nitrogen in the upper atmosphere, and since 1952 from the atmospheric testing of thermonuclear devices. The main source of carbon in groundwater is derived from decaying plants as recharge water moves through the soil zone, with very little coming from dissolved carbon dioxide in rainfall. Carbon-14 concentrations are expressed in 'percentage modern carbon' (pmc). The value of 100 pmc is taken to be the carbon-14 concentration for atmospheric carbon dioxide before 1950. Radiocarbon dating is based on the fact that the carbon in living plants is in near equilibrium with the carbon in the atmosphere and thus the ratio between carbon-14 and stable carbon (carbon-12 and carbon-13) in living plants is the same as in the atmosphere. On dying, the carbon-14 in the organic matter decreases with time due to radioactive decay. The decreasing ratio of carbon-14 to stable carbon is used to date groundwater. In practice, the method is more complicated due to additions of carbon from other sources, such as the dissolution of carbonate rock and the oxidation of ancient organic matter. Various models have been used to take account of reactions between the groundwater and aquifer material all aimed at obtaining an accurate estimate of carbon-14 at zero time. Unlike tritium, the radiocarbon method has enabled the dating of groundwaters up to 40,000 years old in many of the aquifers of North Africa and the Middle East (Wagner and Geyh, 1999).

The stable isotope oxygen-18 amounts to only 0.2 per cent of atmospheric oxygen, and the deuterium–oxygen-18 water molecule in natural waters to less than 1 ppm (parts per million). The concentration of the two isotopes in natural water is dependent on temperature fractionation. As temperature increases the lighter isotopes leave the water surface and the remaining water becomes enriched in the heavier isotopes. This is the basis for the interpretation of hydrogeological problems using stable isotopes. The relationship between deuterium and oxygen-18 in natural waters, known as the meteoric water line (MWL), has been used to provide indications of the origin and age of groundwater. Deviations from the MWL can indicate recharge under different climatic conditions.

Groundwater pollution

Groundwater pollution may be defined as the degradation of natural groundwater quality as a result of human activities. In general, groundwater is slower to pollute than surface water, but once polluted it can take a long time for it to return to its natural condition. Fractured and karstic aquifers are the most susceptible to pollution, as groundwater flow is fast and contaminants suffer little or no attenuation. Intergranular aquifers are slower to pollute, as groundwater flow is slower, except near pumped wells where hydraulic gradients are steeper and velocities greater. A thick unsaturated zone provides good protection against pollution in granular aquifers, but not in fractured and karstic aquifers. Once in the saturated zone, dilution with the resident groundwater, adsorption of contaminants on the surface of dispersed silts or clays and chemical reactions can reduce further the concentration of contaminants. Nevertheless, despite these natural defence mechanisms, human activities in agriculture and industry, and the ever-increasing volume of generated waste, solid and liquid, which in the past had been indiscriminately and without treatment disposed of in landfills, rivers or ponds, have led to the pollution of many aquifers. Groundwater quality has continued to decline despite protection measures and legislation in recent years (Mather *et al.*, 1998; Lerner, 2003).

Pollution can be diffuse or localised (point pollution). In diffuse pollution, pollutants enter the groundwater over large areas and can affect not only groundwater but also those rivers with a large baseflow. Nitrate pollution due to agricultural practices is a good example. Diffuse pollution can also occur from leaking sewers. Past industrial activity in Birmingham, UK, has given rise to pollution by chlorinated solvents of the underlying Sherwood sandstone aquifer (Rivett *et al.*, 2005). Numerous latrines and sanitary absorption pits in the Limassol urban area resulted in widespread nitrate pollution of the gravel aquifer (Hydrogeological Services International, 1998). Point pollution is restricted to a small area, usually of a known location. Hydrocarbon spills from tankers, industrial or mine wastes, landfill sites and petrol stations fall into this category. Groundwater pollution particularly due to industry and mining has been extensive in Europe, the USA and the former Soviet Union in the nineteenth and twentieth centuries. Intensive agriculture in Europe since the Second World War has given rise to widespread groundwater pollution from pesticides, insecticides, and nitrogen and phosphate fertilisers.

Groundwater may also become polluted by overpumping, which may draw saline water into fresh water aquifers. Seawater intrusion of coastal aquifers and upconing of saline water lying below fresh water are the two most common causes. In seawater intrusion, pumping reduces or reverses the natural seaward slope of the water table and allows seawater to move inland. Many of the coastal aquifers of the world and the aquifers of many islands have experienced seawater intrusion. Various measures have been used to control the ingress of seawater, including changes in the pumping regime to form a hydraulic barrier, the construction of subsurface impermeable barriers and artificial recharge (Bear *et al.*, 1999).

Groundwater exploitation

Wells, boreholes and pumps

Where groundwater does not flow naturally at the surface as in springs, it is extracted from aquifers by wells or boreholes. A well is a hole excavated in the earth for the purpose of bringing groundwater to the surface. The excavation may be by hand (hand-dug well) or by machinery. Where drilling machines are employed the water well is known as a borehole, a tubewell, a production well or a production borehole (Misstear *et al.*, 2006). Production boreholes can be simple structures where aquifers consist of hard and stable rock or more complicated structures

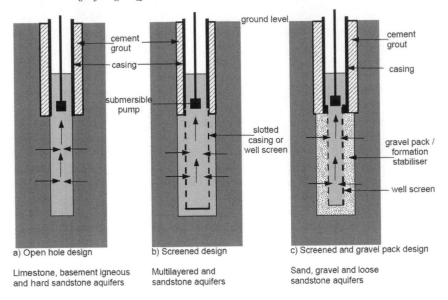

a) Open hole design

Limestone, basement igneous
and hard sandstone aquifers

b) Screened design

Multilayered and
sandstone aquifers

c) Screened and gravel pack design

Sand, gravel and loose
sandstone aquifers

Figure 2.8 Typical production borehole designs for different aquifer lithologies

where aquifers are unconsolidated or semi-consolidated. Typical production borehole designs for different aquifer lithologies are illustrated in Figure 2.8.

In ancient times groundwater was extracted from large diameter wells (generally sufficiently wide to enable a well digger to operate), which were excavated by hand to depths of 5 to 10 metres. Wells are still hand dug today in many countries. Figures 2.9a and 2.9b show photographs of dug wells in the desert of the Taudeni basin in Mauritania and in the Arani-Kortalaiyar basin in south India.

Qanats (also known as *falaj* (plural *aflaj*) in Oman and *karez* in Afghanistan) were another ancient means of tapping groundwater, also employing manual labour. They originated more than 2,500 years ago in Iran and from there spread to the Middle East, Near East and North Africa. The Arabs took *qanats* to Spain, and from there, they were taken by the Spaniards to the New World. A *qanat* is a subterranean tunnel that conveys water by gravity from an underground water source or spring, usually at the mountain foothills or at the edge of alluvial fans, to low-lying areas to be used for drinking and agriculture. Along their length, which could vary from less than 5 to about 30 km, there are vertical shafts spaced 30 to 100 m apart, which were used during construction to remove excavated material and to provide ventilation. *Qanat* systems are still in use, mainly in Iran and and Oman. The use of deep wells and pumps to extract groundwater has caused a lowering of the water table and the drying up of many *qanat* systems in Morocco, Syria, Yemen, Uzbekistan and Iran (Lightfoot, 2003; Motiee *et al.*, 2006). In recent years there has been a renewed interest in *qanats* with the formation of the International Centre for Qanats and Hydraulic Structures in 2003 under the auspices of UNESCO.

Figure 2.9a Dug well in Mauritania for animal watering (photo courtesy B. Burnet, 2007)

Figure 2.9b Large diameter brick-lined well in the Arani-Kortalaiyar basin, south India (photo by the author, 2003)

Figure 2.10 Deep borehole rotary drilling in the Disi Sandstone aquifer, south Jordan (photo by the author, 2011)

Deep wells or boreholes are constructed using drilling machines. Detailed description of drilling methods can be found in Campbell and Lehr (1973), Misstear *et al.* (2006) and Sterrett (2007). There are two main drilling methods: percussion (or cable-tool) and rotary (direct or reverse). In percussion drilling a borehole is advanced by lifting and dropping a cutting tool attached at the lower end of a steel cable (the Chinese used bamboo in ancient times). The cutting tool or bit crushes the rock which is removed by a bailer. In rotary drilling the bit is attached to the lower end of a rotating hollow pipe (drill pipe). Drill cuttings are brought to the surface by circulating fluid (the drilling fluid), which can be water, air or specially made mud from bentonite clay or polymer. The Chinese were the first to use percussion boring about 2,600 years ago to drill brine wells in the Sichuan province. The first drilled well in the USA was in 1859 for oil to a depth of 21m (69 feet) by Colonel Edwin L. Drake (also known as the Drake well). Shortly after, wells drilled to about 300 m depth were put down using rope-suspended iron bars carried by two men. Percussion drilling was gradually replaced by rotary drilling, which enables the drilling of boreholes more speedily to several thousand metres. Modern water wells are on average around 150 m in depth and 250–500 mm in diameter. Much deeper wells, 400–1,000 m, have been constructed in recent years, especially in the Middle East (Figure 2.10). Borehole production rates are variable but generally in the range of 10 to 100 ls^{-1}. However, many of the boreholes exploiting the extensive African basement rock aquifers have yields which are less than 1 ls^{-1}. At the other extreme, boreholes in karstic or strongly fissured aquifers can yield more than 100 ls^{-1}, often with very little drawdown.

Figure 2.11 Groundwater being drawn from a dug well using a Persian wheel, Cyprus 1930s–1940s (photo, courtesy of the Water Development Department)

For many thousands of years the lifting of water from streams or wells was either directly by rope and bucket or using animals (Persian wheel) or the water screw (Archimedes screw). Figure 2.11 shows groundwater being drawn in the 1930s–1940s in Cyprus from a dug well using a Persian wheel driven by a donkey.

It was not until the nineteenth century, when the steam engine and later the combustion engine were invented, that it became possible to devise pumping systems capable of extracting large volumes of groundwater from greater depths. Centrifugal pumps were developed in the nineteenth century and later adapted to installation under water in boreholes. Submersible pumps, as they became known, consist of a series of impellers housed in a bowl assembly which are driven by a line shaft (line shaft pumps) or an electric motor below the bowl assembly (electro-submersible pumps). Submersible pumps are capable of extracting groundwater at high discharge rates of 50 to 100 ls^{-1} from depths of 200 to 300 m. Much of the present groundwater extraction for irrigation, public water supply and industry is from boreholes equipped with submersible pumps. Figure 2.12 shows an irrigation borehole in the Arani-Kortalaiyar basin, south India, equipped with a submersible pump, electricity connection and kiosk housing switchgear.

Global groundwater abstraction

Precise estimates of global groundwater abstraction are difficult, mainly because data are not always available or reliable. Shah *et al.* (2007) suggested that abstraction in 2000 was between 950 to 1,000 $km^3\,a^{-1}$, but Wada *et al.* (2010) estimated a lower figure of 734 ± 82 $km^3\,a^{-1}$, and Giordano (2009) 658 $km^3\,a^{-1}$

Figure 2.12 Equipped irrigation borehole in the Arani-Kortalaiyar basin, south India (photo by the author, 2003)

for the mid-1990s. Although estimates may vary there is no doubt that since the 1950s groundwater abstraction has increased almost tenfold. Asia has been the main abstractor, particularly for irrigation, motivated partly by affordable pumping and drilling technology, partly by government subsidies, and partly due to the convenience of groundwater in comparison to surface water. India, the largest world groundwater user, is a prime example of these influences. Over the years, surface water irrigation has declined giving way to numerous mechanised boreholes, 26–28 million in 2000. In the last 60 years, groundwater abstraction in India has increased twentyfold from about 13 km^3 a^{-1} in 1950 to approximately 210 km^3 a^{-1} in 2000, and it is projected to rise to about 260 km^3 a^{-1} in 2010 (Shah, 2005). Other major users in Asia are China, abstracting in 2000 approximately 111 km^3 a^{-1} (Han *et al.*, 2006) and Pakistan 60 km^3 a^{-1} (Giordano, 2009). Turning to the Americas, Mexico abstracted approximately 30 km^3 in 2011 (National Water Commission of Mexico (Comisión Nacional del Agua CONAGUA), 2011). In the USA, the world's second largest groundwater user, abstraction rose from approximately 47 km^3 in 1950 to 115 km^3 in 1980, remaining at about the same level, on average 109 km^3 a^{-1}, since then (Kenny *et al.*, 2009). In Western Europe groundwater abstraction has decreased marginally from about 45 km^3 in 1950 to 40 km^3 in 2000 (Shah, 2005).

On a global scale, groundwater abstraction is still only a fraction (less than 10 per cent) of worldwide recharge, but this is of course misleading. As already discussed, recharge varies both spatially and temporally, so that in many areas of the world where rainfall is low or negligible deficits have developed. Examples can be found in the Middle East and North Africa, in some of the

Mediterranean countries and islands, Western Australia and northern Chile, and in the western USA. Yemen is a notable example of a country where a nationwide deficit seems to exist. Overabstraction in the Sana'a basin was first highlighted by Charalambous (1982). Since then abstraction has risen from about 40 Mm^3 a^{-1} to about 270 Mm^3 a^{-1}, of which approximately 78 per cent is used for irrigation. The deficit in the Sana'a basin in 2005 was approximately 220 Mm^3 and countrywide 900 Mm^3 (0.9 km^3) (Japan International Cooperation Agency (JICA), 2007). Deficits in some countries, especially in those where fossil groundwater is exploited or where abstraction has persistently been exceeded, have become permanent. In North Africa and the Middle East, approximately 67 per cent of the total groundwater use of approximately 40 km^3 in the 1990s was from fossil groundwater (Foster and Loucks, 2006). In the USA, the great High Plains (Ogallala) aquifer has been exploited since the 1950s. The deficit in 2000 due to overabstraction, mainly for irrigation, was estimated to be 312 km^3 (McGuire, 2007), amounting to an average rate of approximately 6 km^3 a^{-1}. In the large Quaternary aquifer of north China, although annual rainfall is 500–600 mm, excessive exploitation, mainly for irrigation, has led to permanent depletion and ground subsidence. In 2000, groundwater abstraction from the shallow and deep aquifers was approximately 21 km^3, approximately 10 per cent greater than recharge (Han, 2006). Foster and Garduño (2004) estimated that groundwater abstraction from the Hai He basin in 1988 was 27 km^3, exceeding recharge by 8.8 km^3. In the Arani-Kortalaiyar (A-K) groundwater basin in south India, failure of the monsoon rains leads to temporary deficits, particularly where aquifers are thin or have low storage capacity (Charalambous and Garratt, 2009). But persistent overabstraction in the subcontinent has led to large deficits of more than 100 km^3 in 1995 (Postel, 1999). Global estimates of groundwater deficits are probably not very accurate. Wada *et al.* (2010) estimated the total global groundwater depletion for subhumid and arid areas in 1960 to be 126 ($\pm$32) km^3 increasing to 283 ($\pm$40) km^3 in 2000, the latter value corresponding to 2 ($\pm$0.6) per cent of the global yearly groundwater recharge. Postel (1999) suggested a value 200 km^3 for 1995, which is similar to that of Wada *et al.* (2010) for the same year of approximately 210 ($\pm$35) km^3.

Groundwater use

In developing countries, 81 per cent of all water used is for agriculture, 8 per cent for domestic use and 11 per cent for industry whereas in developed countries agriculture takes up only 40 per cent of total use whilst domestic and industrial use is much higher, 16 per cent and 44 per cent, respectively (UNDP, 2006). In Asia and the USA more than 70 per cent of groundwater abstraction is for agriculture. By contrast in the European Union on average 55 per cent of the groundwater is used for domestic supply and only 23 per cent for agriculture with the rest for industry (EEA, 2009). In the southern countries of Europe (Italy, Portugal and Spain) 75–89 per cent of groundwater is used for agriculture (EASAC, 2010).

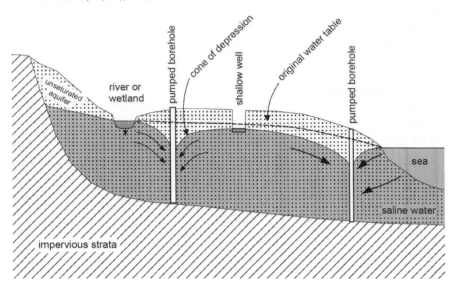

Figure 2.13 Effects of groundwater pumping on rivers, wetlands, shallow wells and coastal aquifers

Impacts due to groundwater abstraction and aquifer management

The effect of groundwater pumping on rivers and wetlands, other users and coastal aquifers is diagrammatically illustrated in Figure 2.13. From the viewpoint of the environment and water resources, the impacts of persistent over-abstraction have been on the whole negative. They include deeper groundwater levels and a decrease in borehole yields, a depletion of available resources, land subsidence, reduction in river flow (baseflow) and damage to wetlands and ecosystems. A further impact is groundwater quality deterioration, especially of coastal aquifers.

On the positive side, the intensive use of groundwater has enabled agricultural growth, provided food, drinking water and sanitation to the rural communities of Africa, Asia, and Central and South America, and enabled the countries of the Middle East and North Africa to meet their ever-increasing demand for drinking water and irrigation. However, groundwater is not inexhaustible and benefits cannot continue forever. In many countries, adverse effects are already evident: groundwater exploitation is becoming unsustainable and uneconomic, while measures to protect aquifers, water resources and the environment are becoming increasingly more costly and difficult to enforce. Land subsidence in many of the large cities of the world (Beijing, Bangkok, Mexico City, San Joaquin and Santa Clara Valleys in the USA, Tokyo and others) has caused damage to infrastructure and buildings, and flooding of coastal areas (Domenico and Schwartz, 1990).

In order to manage groundwater resources sustainably two management concepts have evolved: the concept of 'safe yield' and the concept of 'conjunctive use'. A development from the latter is the relatively recent concept of integrated water resources management (IWRM), which takes a holistic view of water resources that includes land, water and the environment.

The safe yield of an aquifer or a groundwater basin may be defined as the quantity of water that can be abstracted without causing an undesired result. An undesired result could be any of the adverse impacts mentioned above. Because groundwater is generally a renewable resource, safe yield is related to the average annual recharge of the basin or aquifer, such that groundwater extraction should not exceed the annual average recharge. Fossil aquifers, which receive no recharge, present a special case. Their exploitation should be progressive in order to avoid excessive cones of depression, dewatering and premature depletion of groundwater reserves, land subsidence, damage to borehole structures and significant diminution of springs and seepages feeding oases, wetlands, *sebkhas* etc.

The management of a basin by conjunctive use involves the coordinated use of surface water and groundwater resources, thereby increasing the safe yield of a basin, and providing for more flexible management. Groundwater reserves may be enhanced by artificial recharge of surplus surface water into aquifers; or groundwater may be pumped into rivers in the summer to maintain ecological flows or abstractions downstream. The benefits of conjunctive use can be significant, though in practice, to achieve them requires extensive engineering works and facilities involving high costs, experienced personnel, and complicated operational systems.

The Global Water Partnership Technical Advisory Committee (2000) defined IWRM as

> a process which promotes the co-ordinated development and management of water, land and related resources, in order to maximise the resultant economic and social welfare in an equitable manner without compromising the sustainability of vital ecosystems.

The definition is far reaching and implies that activities that may have a direct or indirect impact on water resources should be integrated. However, the IWRM objectives constitute an ideal which in practice is not always easy to achieve. An interesting critique of IWRM and its limited application so far has been presented by Biswas (2008).

Groundwater measurement and allocation

Groundwater is an invisible resource. Therefore, unlike surface water, the direct measurement of its flow is not possible. Instead measurements have to rely on indirect methods. Flow nets based on the Darcy equation provide a means for estimating the groundwater flow in a particular section of an aquifer providing the aquifer transmissivity and hydraulic gradient are known. The estimation of the volume of groundwater stored in an aquifer requires knowledge of storativity and the dimensions (saturated thickness and areal extent) of the aquifer. In unconfined aquifers the stored groundwater is in the void space while in confined aquifers there is in addition the groundwater stored under pressure. However, not all the groundwater stored is recoverable or exploitable. In relatively thin, shallow, unconfined aquifers it may be possible to extract most (70–80 per cent) of the stored volume whereas in aquifers that are several hundred metres thick the exploitable quantity is much less. In deep confined aquifers it may only be possible to extract only the part of the groundwater under pressure without any contribution from the void space. Changes in aquifer storage are reflected in groundwater level fluctuations, which can be obtained from the monitoring of wells or boreholes. Multiplying the change in groundwater level over a given period by the storativity provides an estimate of the change in groundwater storage. A persistent decline in groundwater levels would tend to suggest that extraction exceeds replenishment. Because aquifers are complex and their hydraulic properties are likely to be different over relatively short distances, estimates of flow and storage invariably carry a margin of error, even in areas where the hydrogeology is well known. Comprehensive investigations, often on a regional scale, can provide many of the answers, but there is always the risk of 'diminishing returns' as the costs escalate. Thus, in making groundwater allocations, not only the uncertainty of estimates but also temporal variability have to be borne in mind. In addition, the effects of pumping on other users and the environment need to be assessed. When large quantities of groundwater are to be abstracted, it is good policy to adopt a staged development starting at smaller quantities and progressively increasing to the desired amount, while at the same time monitoring the effects on the resource and hydro-environment by measuring groundwater levels, springflows, pond or lake water levels, streamflows and other water features. Increasingly, groundwater models are employed to estimate groundwater resources and aquifer safe yield, and the impacts of abstraction on river flows, wetlands and other users. A simplified method based on variations in aquifer storage in relation to sustainable yield estimated by groundwater modelling has been used in the Arani-Kortalaiyar basin aquifer in south India (Charalambous and Garratt, 2009).

Concluding remarks

Groundwater, being hidden underground, has always been mysterious, but there has been much progress during the last century in both understanding it better and quantifying its resources. Its storage in aquifers and its slow and diffuse movement through their interstices, distinguishes it from the faster flowing surface and underground streams that follow more or less defined channels. As groundwater is recharged by infiltration of rainfall or surface water, it has a physical link with the land above it, but due to its mobility and hydrogeological complexity the relationship between the parcel of land and the groundwater beneath is not always clear. Unlike streams, aquifers are generally slow to contaminate, but once polluted their cleaning up can be complex, laborious, and may take many years to achieve. Aquifers are often large, measuring from a few to several hundred square kilometres, and can stretch across lands owned by different nations. Most are rechargeable but for a few, known as fossil aquifers, that are found in the arid areas of the world recharge ceased long ago. But even rechargeable aquifers are not inexhaustible, as has been demonstrated by many aquifers of the world, where large deficits have developed. The proper management of both types of aquifers and recognising that groundwater can be scarce, is important in ensuring its sustainability. Groundwater is a common pool resource. When groundwater is extracted from an aquifer at a given location, there is an effect that spreads to other points within it that can have an impact on nearby abstractors, springs, rivers and wetland ecosystems.

In the chapters that follow the legal and economic implications of the various aspects of groundwater highlighted above are addressed, such as its occurrence and mobility, its finite nature, and the impacts of excessive abstractions on water resources and the environment. Transferable groundwater rights as a management tool are discussed once these aspects have been addressed.

References

Allen, R.G., Pereira, L.S., Raes, D. and Smith, M. (1998) *Crop evapotranspiration, guidelines for computing crop water requirements*. FAO Irrigation and Drainage Paper 56. Rome: Food and Agricultural Organisation of the United Nations.

Appelo, C.A.J. and Postma, D. (2005) *Geochemistry, groundwater and pollution*, 2nd edn. Leiden: A.A. Balkema.

Aristotle (1931 [350 BC]) *Meteorologica*, translated by E W Webster. Oxford: Clarendon Press.

Bear, J., Cheng, A.H.-D., Sorek, S., Quazar, D. and Herrera, I. (eds) (1999) *Sea water intrusion in coastal aquifers: concepts, methods and practices*. Dordrecht: Kluwer Academic Publishers.

Biswas, A.K. (2008) Integrated water resources management. Is it working? *International Journal of Water Resources Development*, 24(1): 5–22.

Boulton, N.S. (1963) Analysis of data from non-equilibrium pumping tests allowing for delayed yield from storage, *Proceedings of the Institution of Civil Engineers*, 26: 469–82.

Campbell, M.D. and Lehr, J.H. (1973) *Water well technology*. New York: McGraw-Hill.

Charalambous, A.N. (1982) Problems of groundwater development in the Sana'a basin, Yemen Arab Republic. In G. P. Jones (ed.) *Improvements of methods of long-term prediction of variations in groundwater resources and regimes due to human activity*. Proceedings of the Exeter Symposium, July 1982. IAHS publication 136. Washington, DC: IAHS.

Charalambous, A.N. and Garratt, P. (2009) Recharge-abstraction relationships and sustainable yield in the Arani-Kortalaiyar groundwater basin, India. *Quarterly Journal of Engineering Geology and Hydrogeology*, 42: 39–50.

Chebotarev, I.I. (1955) Metamorphism of natural waters in the crust of weathering, *Geochim. Cosmochim. Acta*, 8(137170): 22–28.

CONAGUA (2011) 2030 Water Agenda, www.conagua.gob.mix.

Cooper, H.H. and Jacob, C.E. (1946) A generalised graphical method for evaluating formation constants and summarising well field history, *American Geophysical Union Transactions*, 27: 526–34.

Cowling, E.B. (1982) Acid precipitation in historical perspective, *Environmental Science and Technology*, 16(2): 110A–21A.

Darcy, H. (1856) *Les fontaines publiques de la Ville de Dijon*. Paris: Dalmont.

Davis, S.N. and DeWiest, J.M. (1966) *Hydrogeology*. New York: John Wiley & Sons.

Döll, P. and Fiedler, K. (2008) Global-scale modelling of groundwater recharge, *Hydrology and Earth System Sciences*, 12: 863–85.

Domenico, P.A. and Schwartz, F.W. (1990) *Physical and Chemical Hydrogeology.* New York: John Wiley & Sons.

Dupuit, J. (1863) *Etudes théoriques et pratiques sur le movement des eaux dans les canaux découverts et a travers les terrains perméables,* 2nd edition. Paris: Dunod.

Dupuy, C.J., Healey, D., Thomas, M., Brown, D., Siniscalchi, A. and Dembek, Z. (1992) A survey of naturally occurring radionnuclides in groundwater in selected bedrock aquifers in Connecticut and implications for public health policy. In C. E. Gilbert and E. J. Calabrese (eds) *Regulating drinking water quality.* Boca Raton, FL: Lewis.

Edmunds, W.M. and Walton, N.R.G. (1980) A geochemical and isotopic approach to recharge evaluation in semi-arid zones – past and present. In *Arid zone hydrology: Investigations with isotope techniques.* Vienna: IAEA.

European Academics Science Advisory Council (EASAC) (2010) *Groundwater in the southern member states of the European Union: An assessment of current knowledge and future prospects,* EASAC policy report 12. Halle: EASAC.

European Commission (1980) Council Directive relating to the quality of water intended for human consumption (80/778/EEC), *Official Journal of European Communities,* L229: 1–8.

European Commission (2000) Water Framework Directive 2000/60/EC of the European Parliament and of the Council of 23 October 2000 establishing a framework for community action in the field of water policy, *Official Journal of the European Communities,* L327, 22/12/2000, pp0001–0073. Also available at http://europa.eu.int/eur-lex/en

European Commission (2001) Commission recommendation of 20 December 2001 on the protection of the public against exposure to radon in drinking water supplies, *Official Journal of European Communities,* L344, 85.

European Environment Agency (EEA) (2009) *Water resources across Europe: Confronting water scarcity and drought,* EEA Report No 2/2009. Copenhagen: EEA.

Fetter, C.W. (2004) Hydrogeology: A short history, Part 2, *Ground Water,* 42(6): 949–53.

Forcheimer, P. (1886) Uber die Ergebikei von Brunen Anlagen und Sickerschlitzen, *Zeitschrift des Archtekte- und Ingenieur Vereins zu Hannover,* 32:539–64.

Foster, S. and Garduño, H. (2004) *China: Towards sustainable groundwater resource use for irrigated agriculture on the north China plain.* Washington, DC: GW-MATE, The World Bank.

Foster, S. and Loucks, D. P. (eds) (2006) *Non-renewable groundwater resources. A guidebook on socially-sustainable management for water-policy makers,* IHP-VI, Series on Groundwater No 10. Paris: United Nations Educational, Scientific and Cultural Organisation (UNESCO), International Association of Hydrogeologists (IAH), World Bank GW-MATE.

Garrels, R.M. and Christ, C.L. (1965) *Solutions, minerals and equilibria.* New York: Harper and Row.

Giordano, M. (2009) Global groundwater? Issues and solutions, *Annual Review of Environment and Resources,* 34: 7.1–7.26. Colombo: International Water Management Institute (IWMI).

Global Water Partnership Technical Advisory Committee (2000) *Integrated water resources management.* TAC Background papers No 4. Stockholm: Global Water Partnership.

Gorrell, H.A. (1958) Classification of formation waters based on sodium chloride content, *American Association of Petroleum Geologists Bulletin,* 46: 1990–2002.

Gray, N. F. (1994) *Drinking water quality problems and solutions.* New York: John Wiley and Sons.

Green, T.R., Taniguchi, M., Kooi, H., Gurdak, J.J., Allen, D.M., Kevin, K.M., Treidel, H. and Aureli, A. (2011) Beneath the surface of global change: Impacts of climate change on groundwater, *Journal of Hydrology,* 405: 532–60.

Grindley, J. (1969) *The calculation of actual evaporation and soil moisture deficits over specified catchment areas,* Hydrological Memoir no38. Bracknell: Meteorological Office.

Halley, E. (1687) An estimate of the quantity of vapour raised out of the sea by the warmth of the sun, *Philosophical Transactions of the Royal Society of London* 16: 366–70.

Han, Z. (2006) Alluvial aquifers in North China Plain, International symposium – Aquifers Systems Management, 30 May–1 June 2006, Dijon, France.

Han, Z., Wang, H. and Chai, R. (2006) *Transboundary aquifers in Asia with special emphasis on China.* Beijing: UNESCO.

Hantush, M.S. (1956) Analysis of data from pumping tests in leaky aquifers, *Transactions of the American Geophysical Union,* 37: 702–14.

Healey, R.W. (2010) *Estimating groundwater recharge.* Cambridge: Cambridge University Press.

Heath, R.C. (2004) *Basic ground-water hydrology,* 10th printing revised, Water Supply Paper 2220. Washington, DC: US Geological Survey.

Hele-Shaw, H.S. (1898) The flow of water, *Nature* 58: 33–6.

Helms, E.G. and Rydell, S. (1992) Regulation of radon in drinking water. In C.E. Gilbert and E.J. Calabrese (eds) *Regulating drinking water quality.* Boca Raton, FL: Lewis.

Hem, D.H. (1985) *Study and interpretation of the chemical characteristics of natural water.* 3rd edn. Water Supply Paper 2254. Washington, DC: United States Geological Survey.

Hiscock, K.M. (2005) *Hydrogeology: Principles and practice.* Leiden: Blackwell Science Ltd.

Hiscock, K.M., Sparkes, R. and Hodgson, A. (2012) Evaluation of future climate impacts on European groundwater resources. In H. Treidel, J.L. Martin-Bordes and J.J. Gurdak (eds) *Climate change effects on groundwater resources. A global synthesis of findings and recommendations.* Oxford: CRC Press.

Homer (1924 [800 BC]) *The Iliad,* translated by A.T. Murray. Loeb Classical Library Volume 1. Cambridge, MA: Harvard University Press.

Hydrogeological Services International (1990) *Hydrogeology of the Disi Sandstone Aquifer, Water Resources Policies and Management,* United Nations Development Programme UNDP/DTCD Project JOR/87/003, The Hashemite Kingdom of Jordan.

Hydrogeological Services International (1998) *Garyllis aquifer study, Limassol, Cyprus.* Unpublished report to the Water Board of Limassol.

JICA (2007) *Study for the water resources management and rural water supply improvement in the Republic of Yemen: Water resources management action plan for Sana'a basin.* Sana'a, The Republic of Yemen: National Water Resources Authority (NWRA) Ministry of Water and Environment.

Karplus, W.J. (1958) *Analog simulations.* New York: McGraw-Hill.

Kazemi, G.A., Lehr, J.H. and Perrochet, P. (2006) *Groundwater age.* Hoboken, NJ: Wiley-Interscience, Wiley and Sons.

Kenny, J.F., Barber, N.L., Hutson, S.S., Linsey, K.S., Lovelace, J.K. and Maupin, M.A. (2009), *Estimated use of water in the United States in 2005,* Circular 1344. Washington, DC: U.S. Department of the Interior, United States Geological Survey.

Kruseman, G.P. and de Ridder, N.A. (1992) *Analysis and evaluation of pumping test data,* 2nd edn, Publication 47. Wageningen: International Institute for Land Reclamation and Improvement.

Lamarck, J.B. (1802) *Hydrogéologie.* Paris: Museum d'Histoire Naturelle (Jardin des Plantes).

Lerner, D.N. (ed.) (2003) *Urban groundwater pollution.* Leiden: A.A. Balkemas.

Lerner, D.N., Issar, A.S. and Simmers, I. (1990) *Groundwater Recharge: A Guide to Understanding and Estimating Natural Recharge.* International Contributions to Hydrogeology Volume 8. Hannover: Verlag Heinz Heise.

Lightfoot, D.R. (2003) Traditional wells as phreatic barometers: A view from qanats and tubewells in developing arid lands. *Proceedings of the Universities Council on Water Resources (UCOWR) Conference: Water security in the 21st century*, Washington DC.

Lloyd, J.W. and Heathcote, J.A. (1985) *Natural inorganic hydrochemistry in relation to groundwater: An introduction.* Oxford: Clarendon Press.

Lohman, S.W. (1972) *Groundwater hydraulics.* U S Geological Survey Professional Paper 708. Washington, DC: United States Government Printing Office.

Lohman, S.W. and others (1972) *Definitions of selected ground-water terms – revisions and conceptual refinements*, US Geological Survey Water Supply Paper 1988. Washington, DC: United States Goverment Printing Office.

Lucas, J. (1877a) The Chalk water system, *Proceedings of the Institution of Civil Engineers*, 47: 70–167.

Lucas, J. (1877b) Hydrogeology: one of the developments of modern practical geology, *Transactions of the Institute of Surveyors* 9: 153–84.

Mariotte, E. (1686) *Traites du mouvement des eaux et des autres corps fluids.* Paris.

Mather, J., Banks, D., Dumpleton, S. and Fermor, M. (eds) (1998) *Groundwater contaminants and their migration*, Special Publication 128. London: Geological Society.

McDonald, M.G., and Harbaugh, A.W. (1988) *A modular three-dimensional finite-difference ground-water flow model*, Techniques of Water-Resources Investigations of the United States Geological Survey, Book 6. Washington, DC: United States Government Printing Office.

McGuire, V.L. (2007) *Changes in water levels and storage in the High Plains aquifer, predevelopment to 2005*, Fact Sheet 2007–3029, Groundwater Resources Program. Denver, CO: United States Department of the Interior, United States Geological Survey.

Mead, D.W. (1919) *Hydrology: the fundamental basis of hydraulic engineering.* New York: McGraw-Hill.

Meinzer, O.E. (1942) Definition of hydrology. In O.E. Meinzer (ed.) *Hydrology*. Physics of the Earth. New York: Dover Publications.

Meinzer, O. E. (1968 [1923]) *Outline of ground-water hydrology with definitions.* United States Geological Survey Water Supply Paper 494. Washington, DC: United States Government Printing Office.

Misstear, B., Banks, D. and Clark, L. (2006) *Water wells and boreholes.* Chichester: John Wiley and Sons.

Monteith, J.L. (1965) Evaporation and environment. In G.E. Fogg (ed.) *The state and movement of water in living organisms*, Symposium of the Society of Experimental Biology 19. New York: Academic Press.

Motiee, H., McBeane, E., Semsar, A., Bahram, G. and Ghomaschi, V. (2006) Assessment of the contributions of traditional qanats in sustainable water resources management, *International Journal of Water Resources Development*, 22(4): 575–88.

Narasimhan, T.N. (1998) Hydraulic characterization of aquifers, reservoir rocks, and soils: A history of ideas, *Water Resources Research*, 34(1): 33–46.

National Research Council (1986) *Global change in the geosphere-biosphere.* Washington, DC: National Academy Press.

National Research Council (1991) *Opportunities in the hydrologic sciences.* Washington, DC: National Academy Press.

Owen, M. (1991) Groundwater abstraction and river flows. *Journal of the Institute of Water and Environmental Management*, 5: 697–702.

Palissy, B. (1580) *Discours admirable de la nature des eaux et fontaines tant naturelles qu'artificielles*, Paris.

Parkhurst, D.L., Plummer, L.N. and Thorstenson, D.C. (1980) *PHREEQE – A computer program for geochemical calculations*, United States Geological Survey Water Resources Investigations Report 80–96. Washington, DC: United States Government Printing Office.

Parkin, G., Birkinshaw, S., Rao, V., Murray, M. and Younger, P.L. (2002) *A numerical modelling approach to the estimation of impact on river flows*. Impact of Groundwater Abstractions on River Flows (IGARF II), R&D Project Record W6-046/PR. Bristol: Environment Agency.

Penman, H.L. (1948) Natural evaporation from open water, bare soil and grass, *Proceedings of the Royal Society, London*, Series A, 193(1032): 120–45.

Perrault, P. (1678) *De l'origine des fontaines*. Paris: Jean et Laurent d'Houry.

Plato (1892 [360 BC) *Dialogues, Critias*, translated by B. Jowett, 3rd edn. Oxford: Oxford University Press.

Plummer, L.N., Prestemon, E.C., and Parkhurst, D.L. (1994) *An interactive code (NETPATH) for modeling NET geochemical reactions along a flow PATH, version 2.0*, U.S. Geological Survey Water-Resources Investigations Report 94-4169. Washington, DC: United States Government Printing Office.

Polubarinova-Kochina, P.Y. (1962) *Theory of ground-water movement*. Princeton, NJ: Princeton University Press.

Postel, S. (1999) *Pillar of sand: Can the irrigation miracle last?* New York: W W Norton Company.

Prickett, T.A. and Lonnquist, C.G. (1968) Comparison between analog and digital simulation techniques for aquifer evaluation. *Use of Analog and Digital Computers in Hyrology Symposium*. Tuscon, AZ: International Association of Scientific Hydrology.

Puri, S. (ed.) (2001), *Internationally (shared) aquifer resources management*, IHP-VI, IHP Non Serial Publications in Hydrology. Paris: UNESCO.

Pyne, R.D.G (2005) *Aquifer storage recovery: a guide to groundwater recharge through wells*, 2nd edn. Gainesville, FL: ASR Publishers.

Rivett, M.O., Shepherd, K A., Keeys, L. and Brennan, A.E. (2005) Chlorinated solvents in the Birmingham aquifer, UK: 1986–2001, *Quarterly Journal of Engineering Geology and Hydrogeology*, 38: 337–50.

Rushton, K.R. and Tomlinson, L.M. (1995) Interaction between rivers and the Nottingham Sherwood Sandstone Aquifer. In P.L. Younger (ed.) *Modelling river-aquifer interactions*. British Hydrological Society Occasional Paper No. 6. London: British Hydrological Society.

Scanlon, B.R., Keese K.E., Flint, A.L., Flint, L.E., Gaye, C B.,W. Michael Edmunds, W.E., and Simmers, I (2006) Global synthesis of groundwater recharge in semiarid, and arid regions, *Hydrological Processes*, 20: 3335–70.

Shah, T. (2005) Groundwater and human development: Challenges and opportunities in livelihoods and environment, *Water, Science & Technology* 51(8): 27–37.

Shah, T., Burke, J., Villholth, K., Angelica, M., Custodio, E., Daibes, F. *et al.* (2007) Groundwater: A global assessment of scale and significance. In D. Molden (ed.) *Water for food, water for life: A comprehensive assessment of water management in agriculture*. London: Earthscan.

Shiklomanov, I.A. and Sokolov, A.A. (1983) Methodological basis of world water balance investigation and computation. In der Beken, A. and Herrmann, A. (eds) *New approaches in water balance computation*, International Association for Hydrological Sciences Publication 148 (Proceedings of Hamburg Symposium). Wallingford: International Association for Hydrological Sciences.

Skeppström, K. and Olofsson, B. (2007) Uranium and radon in groundwater: An overview of the problem, *European Water* 17/18: 51–62.

Sophocleous, M., Koussis, A., Martin, J.L. and Perkins, S.P. (1995) Evaluation of simplified stream-aquifer depletion models for water rights administration. *Groundwater* 33: 579–88.

Sterrett, R.J. (ed.) (2007) *Groundwater and wells*, 3rd edn. Minnesota, MN: Smyth and Company.

Taniguchi, M., William Burnett, W.C.J., Cable, J.E. and Turner, J.V. (2002) Investigation of submarine groundwater discharge, *Hydrological Processes*, 16: 2115–29.

Theis, C.V. (1935) The relation between the lowering of the piezometric surface and the rate and duration of discharge of a well using underground storage, *Transactions of the American Geophysical Union*, 166: 519–24.

Thiem, A. (1887) Verfahress für Naturlicher Grundwassergeschwindegkiten. *Polytechnischer Notizblatt* 42: 229.

Todd, D.K. (1959) *Ground-water hydrology*. 1st edn, New York: John Wiley & Sons.

Todd, D.K. and Mays, L.W. (2005) *Groundwater hydrology*, 3rd edn. New York: John Wiley & Sons.

Treidel, H., Martin-Bordes, J.L. and Gurdak, J.J. (eds) (2012) *Climate change effects on groundwater resources: A global synthesis of finding and recommendations*. Leiden: CRC Press.

UNESCO/WMO (1992) *International glossary of hydrology*. Geneva: World Meteorological Organisation.

United Nations Development Programme (UNDP) (2006) *Human development report, Beyond scarcity: Power, poverty and the global water crisis*. New York: United Nations Development Programme.

United States Environmental Agency (US EPA) (2009) *National primary drinking regulations*. San Francisco, CA: Office of Groundwater and Drinking Water.

Vengosh, A., Irschfeld, D., Vinson, D., Dwyer, D., Raanan, H., Rimawi, O., Al-Zoubi, A., Akkawi, E., Marie, A., Haquin, G., Zaarur, S. and Ganor, J. (2009) High naturally occurring radioactivity in fossil groundwater from the Middle East, *Environmental Science and Technology*, 43(6): 1769–75.

Vitruvius, M.P. (1860 [c. 28 BC]) *De architectura. Book VIII*, translated by Joseph Givilt. London: John Weale.

Wada, Y., van Beek, L.P.H., van Kempen, C.M., Reckman, J.W.T M., Vasak, S. and Bierkens, M.F.P. (2010) Global depletion of groundwater resources, *Geophysical Research Letters*, 37, L20402.

Wagner, W. and Geyh, M.A. (1999) *Application of environmental isotope methods from groundwater studies in the ESCWA region*, Economic and Social Commission for Western Asia, United Nations. Hannover: Bundesanstalt für Geowissenschaften und Rohstoffe (BGR).

WHO (1984) *Guidelines for drinking water quality, Vol. 2. Health criteria and other supporting information*. Geneva: World Health Organisation.

WHO (2011) *Guidelines for drinking-water quality*, 4th edn. Geneva: World Health Organisation.

Younger, P. L. (ed.) (1995) *Modelling river-aquifer interactions*, BHS Occasional Paper No 6, BHS National Meeting with the Geological Society of London. London: British Hydrological Society.

3 Legal aspects of groundwater

This chapter provides an overview of groundwater legislation, and together with Chapter 4 on economics, prepares the ground for the discussion of transferable groundwater rights in Chapters 5 and 6. It has a similar objective to Chapter 2, but from the perspective of a technical or non-legal person who, while not wishing (or indeed needing) to become a legal expert, would still find it useful to become acquainted with the more significant aspects of water law. The chapter begins with a short account of the development of groundwater law. It continues with a brief review of more general water law traditions, starting with the ancient hydraulic civilisations of Egypt and Mesopotamia and concluding with modern water legislative systems. Roman law, civil law and English common law receive particular attention as they have formed the basis of water legislation in many countries. A section follows on international groundwater law, which is still in the process of development. The last section deals with groundwater rights. It includes descriptions of the historic water right doctrines (absolute ownership, riparian and prior appropriation) and of the more recent permit system.

Brief review of water law traditions

Development of groundwater law

Groundwater law has been more sluggish to develop than surface water law, partly because, unlike surface water, groundwater is not visible, and therefore, until recently, not as well understood. However, probably the main reason has been that until the last 50–60 years, its use has been limited. As there was ample groundwater to go round, regulations to manage and control its exploitation were not necessary. The increasing use of submersible pumps and boreholes after the 1950s changed all that. As discussed in Chapter 2, groundwater abstraction quickly soared, bringing with it potential conflicts between users, and adverse impacts on the environment. At the same time, pollution from agriculture and industry added its own adverse effects on groundwater quality. It was, therefore, not too long before it was recognised that old doctrines had to be revisited and a new body of groundwater law (and also environmental law) was needed that could address the new challenges. The modern thinking that emerged was to move away from the traditional doctrine of absolute ownership of groundwater by landowners to user rights allocated by the state. The new system, generally known as the permit system, sought to regulate groundwater use by defining the volume and duration of the groundwater right. Landowners could no longer abstract as much groundwater as they wished without having to worry about the effects of this on their neighbours and on the environment. The permit system has not yet been adopted by all countries. For example, in south Asia, where most of the groundwater abstraction is taking place, countries still retain the old doctrine. Until recently, groundwater rights have remained attached to land. Since the 1980s, in a few jurisdictions, they were assigned property rights of their own. This opened the door to their trading independently of the land parcel on which they occurred.

Water law in the ancient hydraulic civilisations

As far as is known, there were no specific laws governing groundwater in ancient times. Water law related primarily to surface water irrigation. It arose in river valleys where many of the old civilisations flourished: the Tigris and Euphrates in Mesopotamia, the Nile in Egypt, the Indus in India and the Huang Ho in China. Wittfogel (1957) named these 'hydraulic societies' or 'hydraulic civilisations'. Babylonian laws in ancient Mesopotamia were codified by Shulgi, the second king of the Third Dynasty, and these formed the basis of the Hammurabi Code, about 1750 BC. The Hammurabi Code contains a number of articles which directly or indirectly refer to water. Articles 53–56 deal with punishments for those who have been negligent in controlling irrigation waters. Hittite laws, more than 3,000 years ago, contain provisions relating to irrigating land, digging wells and diverting water. In Egypt, the land and water belonged to the pharaoh, who as a living god on Earth, granted their use virtually at will (Caponera, 2003). The Egyptians left almost no written water regulations records, but other records suggest that they must have existed. There was a strong management of the Nile waters, including the measurement of the height of the river water, which started as early as around 3400 BC and continued with small interruptions to Roman times.

Water law in ancient India evolved slowly from custom, religion and written codes (Cullet and Gupta, 2009). In the Indus valley, regulations relating to water were contained in the Laws of Manu (c. 1500 BC). Muslim laws were introduced during the Islamic conquests of northern India after the tenth century and by the Moghuls after the sixteenth century, but had no lasting influence (Siddiqui, 1992). English common law was introduced following the colonisation of India by Britain, also after the sixteenth century (Cullet and Gupta, 2009). Some principles have persisted to modern times, e.g. the right of a landowner to abstract as much groundwater under his land without concern about any adverse impact on his neighbour.

Ancient Chinese water law was based on the Confucian concept expressed by the *li*, or rules of propriety, rituals and customs. In the third century BC, a school of legalists maintained that the law, *fa*, governed human relationships and not the *li*.

The combination of the two concepts during the Han dynasty (200 BC–618 AD) gave birth to the Chinese legal system. In Chinese water law there has never been private ownership of water, and the state, represented by the emperor, was responsible for all water regulating activities (Caponera, 2003).

Ancient Greece

Plato in Book VIII of his *Laws* (347 BC) wrote that 'husbandmen have had of old excellent laws about water', which would suggest that water laws had existed in ancient Greece for some time before (Charalambous, 2011). It seems that spring sources could be privately owned (presumably as part of the land), but the use of the water could be shared, providing that the owner of the spring was not deprived of his water. A landowner was entitled to dig on his own land, but if he could not find water, he had a right to a share of the water from his neighbour's well to satisfy his drinking needs. Plato stressed how easily water can be polluted and the need that it should be protected by law.

Roman law

Roman law has been an important source of law to water resources, which in many respects persists to the present day. A short digression on its history is worthwhile. The development of Roman law covers more than a thousand years from the law of the twelve tables (449 BC) to the *Corpus Juris Civilis* of Emperor Justinian (534 AD). Until the time of Justinian there were two principal sources of law, the imperial constitutions and the classical jurisprudence operating under the law of citations. Justinian, in order to make the law more accessible and responsive to the practical needs of the Empire, appointed, in February 528, a commission of ten jurists led by Tribonian, a jurist in the civil service, and Theophilus, a professor of law at the University of Constantinople, to edit the imperial constitutions as a code. The *Codex Iustiniani* was published in little over a year, in April 529. In December 530, the Emperor ordered the compilation of a digest of the jurisprudence of the great Roman lawyers of the second and third centuries AD. The 'Digest' or 'Pandects' (*Digesta Iustiniani Augusti*) was completed in December 533. While this work was in progress, Tribonian worked on an abridged version of the Digest for the purpose of instruction, which was to replace the classic manual of Gaius. This became known as the 'Institutes of Justinian'. It was published in November 533 and came into effect at the same time as the Digest, on 30 December 533. The April 533 draft of the Code was revised and promulgated in December 534, as the *Codex Repetitae Praelectionis*. This edition of the Code was intended to be final. All subsequent publications were issued as detached constitutions, known as 'novels' (*novella*). The language of the novels was Greek (the language of the Eastern Roman Empire) whilst the Code, the Digest and the Institutes were in Latin. The Justinian compilation or *Corpus Juris Civilis* comprises the twelve books of the Code, the four books of the Institutes, the fifty books of the Digest, and the Novels. The Code is divided into twelve books: Book 1 deals with ecclesiastical law, the sources of the law, and the sources of higher officials. Books 2–8 deal with private law, Book 9 with criminal law, and Books 10–12 with administrative law.

In Roman law the ownership and use of water resources may be divided into three categories (Caponera and Nanni, 2007):

1   Waters common to everybody (*res communis*). In this category were waters in the 'negative community' of things that could not be owned by anyone and could be used by anyone without limit or permission, such as air, the sea and wildlife. Running water (*aqua profluens*) belonged to this category.
2   Public waters (*res publicae*). These waters belonged to the people and included rivers with perennial (permanent) flow. The state had the right to the use of any of the public waters and could grant a right of use to private individuals.
3   Private waters. These were waters which could be privately owned and included streams which do not flow permanently (*flumina torrentia*) and groundwater. The owner of the land has the unrestricted right to their use, independently of the consequences that the use could cause to neighbouring lands. This is illustrated in a passage attributed to Marcellus in the Digest, which states: 'no action, not even the action for fraud, can be brought against a person who, while digging on his own land, diverts his neighbour's water supply' (Reid, 2004; Getzler 2004).

In Roman law, the protection of waters from harmful effects was restricted to the adverse effects of torrential rains, prevention of overflows and protection of aqueducts.

Muslim water law

The Prophet Mohammed laid down three main guiding principles with regard to water:

1   Water is a gift of God. Access to water is the right of the Muslim community based on the principle in the Koran that "We made from water every living thing" (The Holy Koran, XXI, verse 30)
2   Muslims have a common share in grass (herbage), water and fire. In relation to the sale of water, the specific hadith states 'Allah's Messenger forbade the sale of water'.
3   Water should not be wasted, even when performing ablutions before prayer on the bank of a fast-flowing large river.

In the Muslim tradition, under the 'right of thirst', a person has a judicial right to take water to quench his thirst or to water his animals. In customary law, water rights can be either individual or collective or belong to a specific tribe. They may be acquired from the cultivation of land or in proportion to contributions for the construction of wells, *qanats*, canals or other water-bearing structures (Caponera, 2001). The quality and quantity of well waters and other water sources is protected by declaring an area of bordering land as a forbidden area, where the digging of new wells is prohibited. This is similar to the concept of 'protection zones' in modern water law.

Civil law

Civil law, as codified in the French Civil Code (the Code of Napoleon) was promulgated in March 1804 at the behest of Napoleon I. The emphasis of civil water law was on water ownership and use. Legislation on water quality and pollution developed much later. The French Civil Code maintained the distinction of Roman law between public and private waters, except that the distinction was made on the basis of navigability and floatability. As in Roman law, the ownership of public waters (i.e. those waters which were considered to be navigable or floatable) was vested in the state and their use required government permit or authorisation. Private waters (i.e. those waters located below, along or on privately owned land) could be freely utilised, subject to certain limitations of a statutory nature, such as servitudes, rights of way etc. The rights to use private waters, surface or underground, was linked to land ownership which recognised the right of the owner to use at pleasure, without limitation, the water on his land. In French Civil law, groundwaters were classified as private waters. According to Articles 552 and 641, the soil carries with it ownership of what is above and below it and every landowner has the right to use and avail himself of the water of springs rising on his land. However, although a landowner could fully use the spring waters located on his own land, he could do so providing he did not harm the lands of his neighbours. The basic law of 1898 limited this ownership right whenever the spring waters were vital to the population or a nearby community. Getches (2008) noted that the Napoleonic Code contained two facets of modern riparian law: the limitation restricting water rights to riparian landowners, and the requirement that the water be restored to its ordinary course. He also noted that the Code provided that in disputes between riparian landowners, the courts should reconcile the interests of agriculture with property rights, a provision which he considered to be a forerunner of the reasonable use doctrine that was eventually incorporated into the American riparian law.

The French Civil Code found its way into a number of European countries, including Belgium, Spain and Italy and from there to their main colonies in Africa, Latin America and the Far East. It also formed the basis of the Mejelle Ottoman Civil Code.

The Spanish Civil Code influenced water law in Latin America whilst the Napoleonic Code influenced water law in Canada and French African colonies. Scots law inherited the Roman law concept which provides a distinction between running water and standing water. Thus, whilst running waters are common to all men because they can have no bounds, water that is standing and capable of bounds may be appropriated (Ferguson, 1907).

A comprehensive description of the earlier water codes in these European countries, including France, is to be found in the *Water laws in selected European countries* (FAO, 1975). A short description is given below.

The Belgian Civil Code

The Belgian Civil Code, like the French Civil Code, distinguishes between public (navigable and floatable waterways) and private (non-navigable and non-floatable) waters; however, riparian landowners using private (flowing) waters have only right of use and not right of ownership. Ownership of the subsoil, and therefore groundwater, belongs to the landowner. However, this right of ownership is restricted where spring waters are used for the water supply of a community or where the owner of the servient landholding has acquired a right to spring waters by title or prescription.

The Spanish Civil Code

The Spanish Civil Code of 1888 closely follows the form and substance of the Code of Napoleon, particularly with regard to its concept of exclusive land ownership. It recognises two types of water: public and private. Essentially, public waters are all waters found on or under public land and private waters are all waters found on or under private land. Public waters are subject to state control, but are considered common property for essential vital needs, drinking, washing etc. Waters that are not given by law the status of public waters fall into the private dominion of the state, the provinces, the municipalities or the individual. Ownership of springs depends on land ownership. Springs arising on public land are public and springs arising on private land belong to the landowner. Spring waters lose their private status and their use may become public when they enter public land. Similarly, still or stagnant waters follow the legal regime of the land on which they occur. According to the general provisions of the Civil Code, the ownership of the land extends to whatever is below. This gives the landowner full ownership of groundwater extracted from wells on his land. The use of groundwaters within public land is subject to the obtaining of an administrative concession.

The Italian Civil Code

The Italian Civil Code of 1865, unlike the French Code, declared public both navigable and non-navigable streams (*fiumi e torrenti*) and a concession was needed for their use. Riparian landowners could use non-navigable streams. The 1933 basic Water Act (*Testo Unico*), the 1942 Civil Code and special subsidiary legislation did away with the concept of private waters. Instead the term non-public waters was introduced for which there seems to be no clear definition. Public waters form part of the public domain of the state. They comprise rivers, torrents, springs, lakes and extracted groundwater that are or could become fit to satisfy public interest needs. However, ownership of non-extracted groundwater is vested in the owner of the overlying land.

The Mejelle Ottoman Civil Code

The 1870 Mejelle Code was based on the Napoleonic Code where this was in conformity with Muslim law principles. The Code defines water as a non-saleable commodity to which everyone has a right (FAO, 1973). This applies to running water which has not been appropriated, to water in wells dug by unknown persons and to water of the sea and large lakes. Groundwater belongs to the community. Rivers are in the public domain when their beds are not privately owned and privately owned when they flow over private land. The Code, although no longer relevant, played an important role in the codification of the water laws in many of the countries of the Ottoman Empire. In the newly formed Turkish Republic, it was replaced by the 1926 Civil Code.

English common law

English common water law developed primarily in response to competing water rights for access of flowing water. It began in the Middle Ages and continued through to the Industrial Revolution, when the energy from flowing rivers was much in demand to drive mills for fulling, grinding and textile manufacture, and in the mining and metal industries. Bracton, in the Middle Ages, and Blackstone much later during the Industrial Revolution, are two of the main historical sources of English common law. Bracton (c 1210–1268) was an English judge and also a clergyman. He is associated with *De legibus et consuetudinibus Angliae* (*On the laws and customs of England*), which was written in Latin and is one of the oldest studies of English common law. The substance of *De legibus* was drawn from the English courts and its form from Roman law. Blackstone (1723–1780) was an English jurist and professor of law at Oxford. In the period 1765–1769, he published in four volumes his *Commentaries on the laws of England*. The work was written in English and presents a systematic treatment of the common law of England.

With the spread of the British Empire in the seventeenth century, English common law influenced the legal systems of numerous of its former colonies, including: the USA and Canada in North America, Australia and New Zealand, India in Asia, Kenya, South Africa and Zimbabwe in Africa, and the islands of Cyprus and Malta.

Groundwater and surface or subterranean streams

As described in Chapter 2, groundwater moves through the interstices of aquifers in a slow and diffuse manner. It is distinct from surface water streams or water moving in underground channels or subterranean streams, such as those found in the special conditions of karst limestone terrains or in fractured igneous rocks where large conduits have developed. Similarly, common law distinguishes between water moving in defined underground channels or streams, which is treated as surface water, and water percolating through underground strata, which is treated as groundwater. The distinction was clearly drawn by the judges in *Chasemore* v. *Richards* (1839):

> The principles which apply to flowing water in streams or rivers, the right to the flow of which in its natural state is incident to the property through which it passes, are wholly inapplicable to water percolating through underground

strata, which has no certain course, no defined limits, but which oozes through the soil in every direction in which the rain penetrates.

Because the identification of subterranean streams by direct visual inspection is not always possible, their course is usually known or surmised indirectly from observations of surface features (sinkholes or caverns in karst areas), and more recently by means of geophysical surveys. The legal requirement is that the presence of an underground channel need not be shown as a matter of fact, but as a reasonable inference from known facts (*Bleachers Association* v. *Chapel-en-le-Frith*, 1932).

Groundwater ownership and use

Groundwater ownership and use has been the subject of two main court cases: *Acton* v. *Blundell* (1843) and *Mayor of Bradford Corporation* v. *Pickles* (1895). Both reaffirmed the Roman law principle that a proprietor of land has the absolute right to abstract all groundwater moving under his land, irrespective of any impact that this may have on the water supply of his neighbour.

The case of *Acton* v. *Blundell* (1843) involved the cutting off of flow to the cotton mills of the plaintiff due to the construction of mining pits and drains by the defendant. The counsel for the plaintiff invoked the maxim of *sic utere tuo (sic utere tuo ut alienum non laedas*: so use your property as not to injure your neighbour's). Tindal CJ's judgment was based on the concept of absolute ownership, and excluded considerations of injury to one's neighbour:

> ... we think the present case, ... is not to be governed by the law which applies to rivers and flowing streams, but that it rather falls within that principle, which gives to the owner of the soil all that lies beneath his surface; that the land immediately below is his property, whether it is solid rock, or porous ground, or venous earth, or part soil, part water; that the person who owns the surface may dig therein, and apply all that is there found to his own purposes at his free will and pleasure; and that if, in the exercise of such right, he intercepts or drains off water collected from underground springs in his neighbour's well, this inconvenience to his neighbour falls within the description of *damnum absque injuria* (damage done without injury), which cannot become the ground of an action.

In *Bradford Corporation* v. *Pickles* (1895) the defendant, a landowner, sunk a well which diverted groundwater from the plaintiff's source of supply. The plaintiff alleged that the defendant sunk the well so as to extract from the plaintiff a premium price for his land, and not in order to use it for his own purposes; and that the defendant's conduct breached the implicit principle that a nuisance motivated by malice is always wrongful. The House of Lords found against the plaintiff on the basis that property rights cannot be compromised by malicious motive and thereby deprive a landowner from doing 'a thing which every landowner in the country may do with impunity if his motives are good'.

Groundwater pollution

In England, common law has been used by the courts to address the pollution of groundwater percolating in an aquifer as recently as 1994 in the Cambridge Water case. Two earlier Victorian cases, *Rylands* v. *Fletcher* (1868) and *Ballard* v. *Tomlinson* (1885), which had been extensively referred to in the Cambridge Water case, also dealt with groundwater contamination, although the former only indirectly. Common law principles have also been used by US courts for the recovery of damages caused by groundwater contamination (*Branch* v. *Western Petroleum*, 1982 and *State* v. *Ventron Corporation*, 1983).

Rylands v. *Fletcher* involved the escape of water from a reservoir constructed on the defendant's land on behalf of the defendant by independent contractors, which resulted in flooding of the claimant's mine. The case established two principles: (1) a person is responsible for damage to others caused by the escape of substances which he has brought and accumulated on his land; (2) a valid claim requires that the use of land has been 'not natural'. In *Ballard* v. *Tomlinson*, the water of the claimant's well was polluted by sewage and refuse from a drain constructed by the defendant. The court found against the defendant on the basis that the claimant had a right to extract water percolating beneath his land and the defendant had no right to contaminate what the claimant was entitled to get.

The Cambridge Water case and its legal implications have been extensively described in a number of publications (Templeman, 1994; Misstear *et al.*, 1998; Howarth and McGillivray, 2001; Charalambous, 2011). It concerned the pollution by the solvent perchloroethene (PCE) of a Chalk water supply borehole owned by the Cambridge Water Company (CWC) at Sawston Mill, Cambridgeshire, from spillages at the premises of Eastern Counties Leather plc (ECL), a tannery founded in 1879, and located about 1.3 miles (2 km) upgradient from the CWC borehole. The tannery was bought by Hutchings and Harding (HH) in 1976. CWC purchased the borehole in 1977 and started abstracting in 1979. In 1994, CWC took legal action against ECL and HH in nuisance and negligence. The High Court in 1991 rejected CWC's claims on all three counts, but on appeal, ECL was held responsible on the basis of *Ballard* v. *Tomlinson*. This was in turn rejected by the House of Lords in 1993 on the basis that ECL, before 1976, could not have reasonably foreseen the subsequent impairment of the water quality of CWC's borehole.

The application of common law principles to deal with the pollution of groundwater and the environment has increasingly given way to state legislation, both national and European. The legislative path is the approach that the Health and Safety Executive and the UK Environment Agency took with regard to the explosion at the Buncefield Oil Storage and Transfer Depot in Hemel Hempstead, Hertfordshire, in December 2005. The explosion resulted in the Chalk groundwaters of the general area becoming contaminated by PFOS (perfluorooctane sulphonate), a surfactant added to fire-fighting foam to aid it spreading (Buncefield Major Incident Investigation Report, 2008). One of the three charges of the criminal action in 2008 against the five companies involved was causing pollution to groundwater in the vicinity of the plant – the other two related to failures to ensure the health and safety of employees and failure to protect persons not in their employment. In June 2010, following a two-month trial, one of the five companies, Hertfordshire Oil Storage Limited (HOSL), pleaded guilty to causing polluting matter (fuel and firewater chemicals) to enter 'controlled waters', contrary to Section 85(1) and (6) of the Water Resources Act 1991.

The modern approach

In the modern approach, the tendency has been to move away from the traditional doctrines of civil and common law, which have become less responsive to the complex issues of groundwater exploitation, groundwater quality and pollution, and groundwater dependent ecosystems. Overall, the modern approach has been one of greater state control and regulation of water resources. Legislation vesting in the state the ownership or custody of water resources has been introduced in many countries (Salman and Bradlow, 2006). Also, the private ownership of groundwater by the overlying landowner and its use at his discretion without concern to any impacts on neighbours has been steadily eroded and replaced by a system of permits granted by the state. There has also been increasing emphasis on groundwater quality and pollution, which, although it has been slow to evolve, nowadays forms part of most of the modern water and environmental legislation.

The modern approach is probably best reflected in the European Union's (EU) Water Framework Directive (WFD) (Directive 2000/60/EC), which emphasises water quality and sustainability. Specifically with respect to groundwater, the WFD require that

> Member States shall implement the measures necessary to prevent or limit the input of pollutants into groundwater and to prevent the deterioration of the status of all bodies of groundwater

and

> Member States shall protect, enhance and restore all bodies of groundwater, ensure a balance between abstraction and recharge of groundwaters.

In December 2006, the second Groundwater Directive (2006/118/EC) was adopted to complement the WFD. The Directive sets out various requirements and targets that need to be met with regard to groundwater quality and pollution. The EU has been active in promoting the application of the WFD not only within the Union, but also outside, especially in Eastern Europe and the former Soviet Union republics of Central Asia.

In England and Wales, there have been various water acts since 1945 (Water Resources Act, 1963; Water Resources Act, 1991; Water Act 2003) aimed at protecting and conserving water resources, and preventing the discharge of polluting substances into underground strata. In recent years, water legislation has been influenced by European law. In the USA, statutes have been introduced to safeguard against overabstraction, prevent adverse impacts to other users and the environment, and ensure that water is used beneficially and not contrary to the public interest. Nevertheless, water law statutes are still influenced by traditional doctrines, riparian in the East and prior appropriation in the West (Tarlock *et al.*, 2002), whilst the absolute ownership doctrine still remains the law influencing groundwater abstraction in Texas. Groundwater protection has relied upon environmental laws or other indirect legislation, such as: the 1972 Clean Water Act, the 1976 Solid Waste Act, the 1980 Comprehensive Environmental Response and Liability Act (CERCLA), also known as the 'Superfund', and the 2006 National Drinking Water Regulation, or 'Ground Water Rule' (GWR). Nevertheless, common law tort remedies are still being used to address aquifer pollution. At state level, there have been difficulties to administer and enforce water quality and groundwater protection laws alongside allocation laws, partly due to different agencies being responsible for regulating groundwater allocation and polluting activities and partly because of concerns that pollution laws would erode water rights (Getches *et al.*, 2002).

Muslim countries, many of which have acquired their independence in the twentient century, have promulgated water laws entrusting central government with the conservation, development and use of water resources (FAO, 1973). In the former Soviet Union and Eastern Europe, the changes from socialist to market-based economies in the 1990s have been accompanied by changes in water legislation (Hodgson, 2006), which in many cases aim towards a convergence with European Community legislation. This process is still in progress. Other countries, particularly from Eastern Europe, which have recently joined the European Union, have had to implement water legislation consistent with the European Water Framework Directive, whilst countries which are not part of the European Union but bordering European Union countries, needed to harmonise their legislation with the European Framework Directive.

In South Africa, with the fall of the apartheid regime in 1994, water legislation that favoured the white community has been replaced by legislation that aims at water sustainability and equity.

The concept of integrated water resources management (IWRM) is also part of the new thinking. IWRM has been briefly discussed in Chapter 2. IWRM principles are increasingly reflected in modern national water legislation and policies (Salman and Bradlow, 2006), including the EU's WFD, which encourages member and non-member states to consider water resources in an integrated manner. There has been endorsement of IWRM during various international conferences, including the Dublin Conference on Water and Environment, 1992, the Rio (Earth Summit) United Nations Conference on Environment and Development (UNCED) Agenda 21, 1992, and the World Summit on Sustainable Development (WSSD) in Johannesburg, 2002. Nevertheless, the ambitious goals of IWRM still remain elusive (Biswas, 2008), and in the case of groundwater basins, which are three-dimensional structures that often consist of complex aquifer systems with slow temporal responses, are likely to be difficult to achieve (Quevauviller, 2007).

International groundwater law

There has been interest in international groundwater law for many years. A compilation of treaties and other legal instruments under international groundwater law was prepared by Teclaff and Utton (1981), which was recently updated by the FAO Legal Office in Rome (Burchi and Mechlem, 2005). However, only a few treaties deal directly with groundwater. Most deal with water resources in general, surface waters or the environment.

In 1986, The International Law Association (ILA) prepared the Seoul Rules on International Groundwaters (ILA, 1986), which extended the 1966 ILA Helsinki Rules to waters in international aquifers. In 2004, the ILA prepared the Berlin Rules on Water Resources (ILA, 2004), which included groundwater and transboundary aquifers. In 1989, a group of scholars specialising in transboundary groundwaters prepared the Bellagio Agreement (Hayton and Utton, 1989). The purpose of the Agreement was to serve as a model international treaty. None of these rules or agreements has any legally binding status. An important development was the 1997 United Nations Convention on the Law of 'Non-Navigational Uses of International Watercourses' (UN, 1997). The Convention, which is yet to be ratified, focused on surface water, but lacks clarity with regard to groundwater (Mechlem, 2003). In the European Union, the WFD requires of member states to cooperate among themselves and with non-member states where the transboundary effects from the use of water within a river basin may extend beyond the Union's boundaries.

In the last few years, there has been a renewed effort by the United Nations to address transboundary groundwaters. A Special Rapporteur was appointed in 2002 by the International Law Commission (ILC) to report on shared natural resources, which, initially, in addition to groundwater, included oil and natural gas. Between 2002 and 2008 the Special Rapporteur submitted five reports. In his Third Report (Yamada, 2007), he provided a draft convention on the law of transboundary aquifers, which he amended in his Fifth Report (Yamada, 2008) following comments and observation from governments. Using the Fifth Report as a basis, the ILC finalised a set of 19 draft articles on the law of transboundary aquifers. The draft law is divided into four parts (ILC, 2008), as follows: Part I 'Introduction' sets out the scope of the proposed law and use of terms; Part II 'General Principles' is concerned with sovereignty of aquifer states, equitable and reasonable utilisation, factors relevant to equitable and reasonable utilisation, obligation not to cause significant harm to other aquifer states, general obligation to cooperate, regular exchange of data and information, and bilateral and regional agreements and arrangements; Part III 'Protection, Preservation and Management' deals with the protection and preservation of ecosystems, recharge and discharge zones, prevention, reduction and control of pollution, monitoring, management and planned activities; and Part IV 'Miscellaneous Provisions' discusses technical cooperation with developing states, emergency situations, protection in time of armed conflict, and of data and information vital to national defence or security. The draft articles were appended to the UN General Assembly resolution A/RES/63/124 during the UN's 63rd

General Assembly in 2008 (UNGA, 2008). The resolution recommended to the states concerned that they should make appropriate bilateral or regional arrangements for the proper management of their transboundary aquifers on the basis of the principles enunciated in the draft articles; and also to consider at a later stage the elaboration of a convention on the basis of the draft articles. A further recommendation was to consider the final form of draft articles in its 66th convention in the autumn of 2011. At its 66th session in October–November 2011, the UN General Assembly did not arrive at a final form of transboundary groundwater law, but rather it reiterated the recommendations made at its 63rd session, and furthermore decided to:

> include in the provisional agenda of its sixty-eighth session the item entitled 'the law of transboundary aquifers' and, in the light of written comments of Governments, as well as views expressed in the debates of the Sixth Committee held at its sixty-third and sixty-sixth sessions, to continue to examine, *inter alia*, the question of the final form that might be given to the draft articles.
>
> (UNGA, 2011)

It seems that the draft articles of the law of transboundary aquifers are a long way from becoming formalised, which given the history of the UN Convention on the Law of the Non-Navigational Uses of International Water Courses, is not surprising. In general, commentators have welcomed the draft articles but expressed disappointment with regard to the inclusion of article 3 on the sovereignty of each state over the portion of the transboundary aquifer that occurs within its territory (McCaffrey, 2011; Dellapenna, 2011). The concept of sovereignty of aquifer states has been considered to be questionable in international law, lacking clarity in the light of the requirement of article 5 for equitable and reasonable utilisation and not likely to protect transboundary aquifers from overexploitation by one or the other of aquifer states. The relevance and potential role of the draft articles in guiding European states in the management of their shared aquifers have been explored by Allan *et al.* (2011). They concluded that the draft articles may offer a common platform and serve as a model for agreements on transboundary aquifers between European Union member states and between European Union states and non-member states.

Groundwater rights

In this book, a groundwater right is considered to be a legal right to abstract and use groundwater, the right having been created by a country's formal legal system. This definition is similar to that of Hodgson (2006) for a water right. Meinzen-Dick *et al.* (2004) suggested a broader definition of water rights, taken from Wiber (1992):

> claims to use or control water by an individual or group that are recognised as legitimate by a larger collectivity than the claimants and that are protected through law.

Also, of some interest is the relatively recent concept of 'human right to water', which entitles everyone to sufficient, safe, acceptable, physically accessible and affordable water for personal domestic uses (General Comment No 15 of the United Nations Committee on Economic, Social and Cultural Rights) (UNCESC, 2003). As the Committee has no law-making powers, this declaration alone cannot create a right to water. A detailed discussion on the human right to water has been given by McCaffrey (2005).

Historically, groundwater rights have been allocated under the absolute ownership doctrine, also known as the English doctrine or rule, and in Texas, as the 'rule or law of capture'. Two other historical doctrines, although applied mainly to surface water, are the riparian doctrine and the prior appropriation doctrine. Permit systems are a relatively recent development. Depending on the jurisdiction, water rights in permit systems take the form of licences, entitlements or concessions issued by the state and, usually, subject to conditions of duration, quantity and type of use. Permit systems have retained some of the features of the historical doctrines. They are also usufructuary, meaning that a landowner has the right to use the water that occurs under or over his land and derive all benefits from that use, but he does not own it.

The riparian and absolute ownership doctrines are fundamentally land based, which means that water rights cannot be transferred independently of land. Recently, in a few countries, the land–water link has been severed and water rights may be transferred or traded on their own. Transferable groundwater rights, the main subject of this book, are discussed at length in the chapters that follow.

Guidelines in preparing national water resources regulations with examples from different countries is found in Burchi and D'Andrea (2003) and Salman and Bradlow (2006). Hodgson (2006) provides a comprehensive review of the theory and practice of modern water rights.

The absolute ownership doctrine

The doctrine gives the right to a landowner to unlimited use of the groundwater under his land irrespective of the consequences to his neighbours. It has its origin in Roman law, civil law and English common law, all of which vest ownership of groundwater in the owner of the land above it. English common law allows the landowner to take as much water as he likes without concern for any effects on his neighbour (*Bradford* v. *Pickles*, 1895). In the US case of *Roath* v. *Driscoll* (1850) it was suggested that another reason for applying this doctrine was the mysterious character of groundwater. At that time, the very limited knowledge of hydrogeology made it difficult to establish a causal connection between withdrawals by a defendant and harm to a plaintiff. Also, in *Frazer* v. *Brown* (1861) the Ohio Supreme Court provided the following colourful explanation in support of the doctrine:

> Because the existence, origin, movement, and course of such waters, [meaning groundwater] and the causes which govern and direct their movements, are

so secret, occult, and concealed that an attempt to administer any set of legal rules in respect to them would be involved in hopeless uncertainty, and would, therefore, be practically impossible.

As already mentioned in this chapter, similar ideas were expressed earlier in the English courts (*Chasemore* v. *Richards*, 1839).

As would be expected, the absolute ownership doctrine potentially leads to overabstraction, and the consequent adverse impacts on other users and the environment. In the Indian subcontinent, the doctrine has allowed smallholders to abstract large quantities of groundwater for irrigation, which, as already mentioned in Chapter 2, has caused overdrafts and declining groundwater levels. In the USA, the doctrines of reasonable use and correlative rights were developed in order to control overabstraction.

The absolute ownership doctrine has served to promote the development of groundwater for municipal supplies. However, because of their large scale, municipal abstractions have invariably had an influence well outside the boundaries of the water undertaker's land. With the introduction of correlative rights, water undertakers run the risk of being taken to court by surrounding users whose supplies were affected. In some states of the USA, public water supply undertakers were given the power of eminent domain over water and, as a result, injuries to overlying landowners were confined to damage remedies.

The absolute ownership doctrine is steadily being abandoned in favour of the modern permit systems of allocation. In the USA, it is still valid in Maine and Texas. In Texas, the 1997 Edwards Aquifer Act has failed to modify the Absolute Ownership Doctrine. (For a discussion of the Edwards aquifer, see Chapter 5). In the important case of *Edwards Aquifer Authority* v. *Day* (2012), the Supreme Court of Texas ruled that land ownership includes an interest in groundwater in place that cannot be taken for public use without adequate compensation guaranteed by article I, section 17(a) of the Texas Constitution ('No person's property shall be taken, damaged, or destroyed for or applied to public use without adequate compensation being made.'). With regard to groundwater ownership the court held that:

> Whether groundwater can be owned in place is an issue we have never decided. But we held long ago that oil and gas are owned in place, and we find no reason to treat groundwater differently.

And also,

> a landowner has a right to exclude others from groundwater beneath his property, but one that cannot be used to prevent ordinary drainage.

The implication of the judgment is to reassert private ownership of the groundwater beneath a landowner's land. And although the absolute ownership doctrine permits that groundwater beneath a landowner's land may be drained by a neighbour's pumping in his own land, the removal of groundwater by a neighbour pumping from a well (such as a deviated well or adit) on his property but tapping the aquifer beneath the landowner's land is not permitted. The state may regulate groundwater production, but it must do so in a way that does not deprive a landowner of his property rights. Historical use is not sufficient grounds to establish groundwater production.

Reasonable use and correlative rights doctrines

As indicated above, the two doctrines were introduced in the USA in an effort to curb the excesses of the absolute ownership doctrine. Under both doctrines, the water right holder still needs to have ownership of the overlying land. However, under the reasonable use doctrine, there is a restriction with regard to the place of use (although, some states allow use on non-overlying land, providing such use does not interfere unreasonably with neighbouring landowners), but little restriction on the nature and amount of use on overlying land. Under the correlative rights doctrine, owners of land overlying an aquifer are each limited to a reasonable share of the total resource (Getches, 2008). This is usually based on the area of land ownership overlying an aquifer. Preference of use is given to owners of overlying land. Users who do not own the overlying land are treated under the prior appropriation doctrine. Appropriative rights are subordinate to correlative rights. Non-overlying use is permitted only if surplus of water is available or water is not needed by overlying landowners.

The Riparian water rights doctrine

The riparian doctrine (from the latin *riparius* meaning 'of a river bank') applies to surface water and, it is therefore mentioned only briefly here. It concerns the water rights of owners of land abutting surface watercourses. Its fundamental principle is that the owner of land bordering a watercourse acquires certain rights to use the water, among which is the right to the flow of the watercourse. Riparian water right principles apply to underground watercourses that occur under a riparian owner's land, provided there is evidence to show that the water in the watercourse flows in a defined channel. However, they do not apply to groundwater under a riparian's land, as made clear in *Chasemore* v. *Richards* (1859), unless it can be demonstrated that a connection exists between surface water and groundwater.

Over the centuries the riparian doctrine has undergone changes. In early times the use of a stream from time immemorial ('ancient use') gave the right to a riparian owner to continue making use of it, even if the use deprived others of the natural flow of the stream. In the 1800s, two concepts emerged, that of 'natural flow' and that of 'reasonable use'. Under the first, every riparian landowner has an equal right to use water in a stream but without causing a diminution of its natural flow. This principle was asserted in the English courts in *Wright* v. *Howard* (1823) and reaffirmed in *Mason* v. *Hill* (1832). Under the second, each riparian landowner has a right to make all reasonable uses of the flow of a stream provided that the use does not affect the reasonable uses of other riparians. The reasonable use doctrine was enunciated in American courts by Justice Story in the case of *Tyler* v. *Wilkinson* (1827).

The prior appropriation doctrine

The prior appropriation doctrine concerns the act of diverting water from a natural source for beneficial use. Prior appropriation refers to the priority of right given to the earliest users. The presumption is that the right to divert water is a 'usufructuary' right and not a 'possessory' right.

Concepts of prior appropriation for flowing waters appear to have existed early in English common law (Getzler, 2004). However, prior appropriation is essentially an American doctrine, applied to the arid lands of the western USA, where the riparian doctrine was found to be constraining to the settlers' mining and agricultural activities. The main thrust towards the development of appropriative water rights started with the California gold rush, following the discovery of gold in January 1848 in the foothills of the Sierra Nevada (Tarlock *et al.*, 2002). There were a number of factors that influenced the process. These included the dryness of the climate and the ephemeral flows of streams, and the fact that mining operations were often at some distance from watercourses necessitating the conveyance of water away from riparian lands. Also, the construction of canals and ditches was a laborious and costly operation, requiring security of investment that only undisputed water rights could provide. Finally, water was essential to mining operations and was needed in large quantities. Underlying these factors was the lack of law and organised government in the early years. Water rights were, therefore, developed on public lands by miners, who were essentially trespassers, and who made their own 'customary' laws of acquisition, priority, use and loss of water rights. Legislation recognising the rights of the appropriators came later in the form of three Acts: the 1866 Mining Act, the 1870 Amendment to Mining Act, and the 1877 Desert Land Act.

The principle of priority is fundamental to the prior appropriation doctrine. The person who makes the first appropriation has the highest priority ('first in time, first in right') and a superior right to later appropriators. In times of scarcity, the senior appropriators are served first. The system has the advantage of providing security, and therefore an incentive for development to senior appropriators, but can stifle efforts to put water to better economic uses or prevent overexploitation. Thus, in the western USA where the appropriation doctrine is still to be found in the modern legislation of permit type systems, it has been tempered by statutory limits on pumping, considerations of impairment of the rights of existing users by taking account of available resources, controlled abstraction of non-rechargeable or slowly rechargeable aquifers, and controlling of groundwater contamination (Getches, 2008). In England and Wales, the 1963 Water Resources Act provided a 'protected right to abstract water' to persons who already had a licence to abstract (Water Resources Act, 1963). However, protected rights have been gradually eroded in subsequent legislation (Water Act, 2003).

Permit systems of water rights allocation

As demand for water grew, the historical water rights doctrines could not easily address the requirements of numerous users, including public supplies to cities, while at the same time protecting water resources from overexploitation and the environment from adverse impacts. They also became difficult to administer, often leading to protracted disputes between individuals, lengthy court cases and piecemeal legislative reforms. For example in England, between 1800 and 1947, there were some 4,500 private and local Acts of Parliament giving rights to use water (Hodgson, 2006). In the eastern USA, the riparian doctrine no longer served the public interest in safeguarding a reliable supply to water (Getches, 2008), whilst in Australia it impeded the development of mining, agriculture, and towns during the nineteenth century (Musgrave, 2011). In the western USA, the prior appropriation and absolute ownership doctrines resulted in the depletion of many of the aquifers and ground subsidence, with the result that statutory limits on pumping had to be imposed in many of the western states.

Other drivers to change were social, as in South Africa, where riparianism was transformed into a damaging regime serving the interests of the white landowning minority (Allan, 2003). There were also political drivers, as in the former socialist Eastern European countries and the former Soviet Union where state control did not create substantive water rights (Hodgson, 2006), and economic, as in Chile, although here political dogma played a dominant role when the Pinochet right-wing regime, which replaced the socialist government of Allende, removed state regulation of water resources and shifted water rights to private ownership (Bauer, 2004).

Permits to abstract water are determined and issued by the competent authority charged with this task by government. Factors that influence the issue of abstraction permits are: status of the resource in terms of availability, impact on existing users, impact on the environment (wetlands, river flows, national and international heritage sites), types of use (irrigation, public supply, industrial) etc. It has become increasingly the case that applicants have to carry out extensive investigations including long-duration pumping of boreholes and attendant monitoring of hydrological features and of other wells in order to determine, usually in the form of an environmental assessment evaluation, potential impacts and how these may be mitigated. In most jurisdictions, permits are not required for water to be used for domestic purposes, normally the upper limit being set at 10–20 $m^3 day^{-1}$. Permits, nowadays, are rarely unconditional. Two of the main conditions are volume to be abstracted and the duration of the permit. A third condition is the payment of abstraction fees or charges to the relevant government department.

The volume to be abstracted is usually defined in the form of annual, daily and hourly rates. The annual volume is normally based on the availability of the resource after taking account of a number of factors:

1   existing exploitation;
2   the protection of existing licensed allocations which permit holders are legally entitled to expect;

3   environmental requirements, such as, groundwater outflows to wetlands necessary to protect habitats;

4   the maintenance of the flow of springs and the baseflow and water level in rivers;

5   the protection of aquifers from sea water contamination in coastal areas; and

6   in some areas, the protection from ground subsidence, which can arise mainly in confined aquifers as a result of over-abstraction.

The available groundwater resource is generally taken to be the safe yield of the aquifer less the various allocations indicated above. In fossil aquifers, where abstraction is from groundwater storage, other management rules may apply, such as, economic limits to extraction (i.e. pumping depths), environmental impacts on oases and *sebkhas*, maintaining the resource for use by future generations and social impacts on dispersed or pastoral communities.

In England and Wales, a permit system was first introduced in the 1963 Water Resources Act. In 1999, the government proposed the development of catchment abstraction management strategies (CAMS) (DETR and Welsh Office, 1999). CAMS had a number of objectives, including the provision of a consistent and structured approach to local water resources management, which recognised both the reasonable needs for water of abstractors and environmental needs, and interestingly, the facilitating of water licence trading (EA, 2001). CAMS introduced the concept of 'resource availability status' as an indicator of the relative balance between committed and available resources in a given catchment. One of the initial outcomes of CAMS was to highlight data gaps in many of the catchments and the need for further studies, particularly in groundwater, in order to enable more robust estimates of resources to be made.

In the past, it has been the case that permits to abstract water were typically of unlimited duration. This is still the case in Chile, the western USA (although, in both, in Chile only recently, the right can be lost if not used, or if water is not used beneficially) and in some states of Australia, although licences may be subject to a periodic review. The tendency in recent years has been towards time-limited permits. In England and Wales the time limit is usually 12 years, but permits can be revoked or varied if not used for 4 years (Water Act, 2003); in Mexico the limit is from 5 to 30 years, but permits may be cancelled if not used for 3 consecutive years (National Water Law 1992, reformed 2004); in Queensland, Australia permits are for up to 10 years (Water Act, 2000) and in New South Wales for 10–20 years (Water Management Act, 2010). In South Africa the maximum duration is 40 years, but subject to review every 5 years (National Water Act 36, 1998). The main reason for adopting time-limited water rights lies in the uncertainties that surround water resources evaluations. These are compounded by climate change, and the impact of abstractions on the environment that can take many years to be identified and resolved. On the other hand, users, especially large water undertakers with statutory responsibilities to provide water to the public (but also requiring a return on their investments), need to be assured of the security of their right in the long term. In the end, the duration of the permit is a balance between these two main issues. In general, the policy in most jurisdictions is that a permit is renewable providing that the need remains justifiable, water is used efficiently and the abstraction does not affect the sustainability of the resource and the environment (Charalambous, 2011).

Fees or charges are applied, usually annually, to abstraction permits exceeding a certain quantity of water. Charges are calculated on the basis of type of use, time of use and the time of the year. Charges are highest where water is used for spray irrigation which has a high water loss. Summer use normally attracts higher charges than winter use.

As already indicated, in permit systems water rights are usufructuary, and the ownership of the resource is vested in the state. In many developing countries there has been resistance by farmers to the concept of ownership or custodianship of groundwater by the state (Salman and Bradlow, 2006). There have also been challenges and compensation claims in Arizona and New Mexico in the USA, and in Spain, in response to legislation vesting groundwater resources in the state and divesting private water rights from landowners and well owners. Both the US and Spanish courts upheld the new legislation, mainly on the grounds that there was a superior common good to be served by vesting groundwater ownership in the state, and that reasonable measures had been provided for in the legislation to mitigate any adverse impacts (Burchi, 1999). However, the private ownership of groundwater still remains a thorny issue, as demonstrated in the US case of *Edwards Aquifer Authority* v. *Day* case discussed above.

Concluding remarks

There has been considerable development in water law in recent years, and with aquifers becoming increasingly depleted, groundwater law has been receiving more attention. Many countries are now revising their water legislation in response to the new challenges of scarcity, uncertainties related to climate change, pollution and environmental impacts. There is a steady movement away from the old doctrines of riparianism, prior appropriation and absolute ownership in favour of permit type systems, which provide a more formal and explicit definition of groundwater rights, particularly in terms of the measure of the right and its duration. The connection of land rights and groundwater rights still persists in most modern water codes, although in the last 20–30 years, a few countries have divorced the two. This in effect makes groundwater rights transferable and capable of being treated as a private good. The significance of this for groundwater resources governance is discussed at length in the chapters that follow.

Legislation to protect groundwater quality from pollution has been, until recently, slow to evolve. However, it is now considered as important as quantity, and the last 30 years have witnessed the emergence of a considerable body of legislation on water quality standards and regulations to protect groundwater from polluting activities.

References

Allan, A. (2003) Comparison between the Water Law Reforms in South Africa and Scotland: Can a generic national law model be developed from these examples? *Natural Resources Journal*, 43(2) 419–89.

Allan, A., Loures, F. and Tignino, M. (2011) The role and relevance of the draft articles on the Law of Transboundary Aquifers in the European context, *Journal of European, Environmental & Planning Law*, 8(3): 231–51.

Bauer, C.J. (2004) *Siren song: Chilean Water Law as a model for international reform.* Washington, DC: Resources for the Future.

Biswas, A.K. (2008) Integrated water resources management: Is it working? *International Journal of Water Resources Development*, 24(1): 5–22.

Buncefield Major Incident Investigation Board (2008) *The final report of the Major Incident Investigation Board – Volumes 1, 2a and 2b.* http://www.buncefieldinvestigation.gov.uk.

Burchi, S. (1999) National regulations for groundwater: Options, issues and best practices. In S.M.A. Salman (ed.) *Groundwater legal and policy perspectives: Proceedings of a World Bank seminar*, World Bank Technical Paper no 456. Washington, DC: World Bank.

Burchi, S. and D'Andrea, A. (2003) *Preparing national regulations for water resources management: principles and practice.* FAO Legislative Study 80. Rome: Food and Agricultural Organisation.

Burchi, S. and Mechlem, K. (2005) *Groundwater in international law. Compilation of Treaties and other legal instruments.* FAO Legislative Study 86. Rome: Food and Agricultural Organisation.

Caponera, D.A. (2001) Ownership and transfer of water and land tenure in Islam. In N.I Faruqui, A.K. Biswas and M.J. Bino (eds) *Water management in Islam.* Tokyo: International Development Research Centre, United Nations University Press.

Caponera, D.A. (2003) Water laws in ancient hydraulic civilisations. In *National and international law and administration: Selected writings.* International and National Water Law Policy. The Hague: Kluwer Law International.

Caponera, D.A. and Nanni, M. (2007) *Principles of water law and administration national and international*, 2nd edn. London: Taylor and Francis.

Charalambous, A.N. (2011) Groundwater and the law, *Quarterly Journal of Engineering Geology and Hydrogeology*, 44: 147–158.

Cullet, P. and Gupta, J. (2009) India: Evolution of water law and policy. In J.W. Dellapenna, and J. Gupta (eds) *The evolution of the law and politics of water.* Dordrecht: Springer.

Dellapenna, J.W. (2011) The customary law applicable to internationally shared groundwater. In G. Eckstein (ed.) Special Issue: Strengthening cooperation on transboundary groundwater resources. *Water International*, 36(5): 584–94.

DETR and Welsh Office (1999) *Taking water responsibly: Government decisions following consultations on changes to the water abstraction licensing system in England and Wales.* London: DETR.

EA (2001) *Managing water abstraction: The catchment abstraction management strategy process.* London: Environment Agency of England and Wales.

EC (2000) Establishing a framework for Community action in the field of water policy. Directive 2000/60/EC of the European Parliament and of the Council of 23 October 2000, *Official Journal of the European Union* L327, 22/12/2000 P001–0073.

EC (2006) On the protection of groundwater against pollution and deterioration. Directive 2006/118/EC of the European Parliament and of the Council of 12 December 2006, *Official Journal of the European Union* L 372/19, 27.12.2006.

FAO (1973) *Water laws in Moslem countries,* Irrigation and Drainage paper 20/1. Rome: Food and Agricultural Organisation.

FAO (1975) *Water laws in selected European countries.* FAO Legislative Study no 10. Rome: Food and Agricultural Organisation.

Ferguson, J. (1907) *The law of water and water rights in Scotland.* Edinburgh: William Green and Sons.

Getches, D.H. (2008) *Water law in a nutshell: Overview and introduction to water law.* 4th edn, St Paul, MN:Thomson-West.

Getches, D.H., MacDonell, J. and Rice, T.A. (2002) Controlling water use: The unfinished business of water quality protection. In A.D. Tarlock, J.N. Corbridge and D.H. Getches (eds) *Water resources management: A casebook in law and public policy,* 5th edn. New York: Foundation Press.

Getzler, J. (2004). *A history of water rights at common law.* Oxford: Oxford University Press.

Hayton, R.D. and Utton, A.E. (1989) Transboundary groundwaters: The Bellagio Treaty, *Natural Resources Journal,* 29: 663–720.

Hodgson, S. (2006) *Modern water rights theory and practice,* FAO Legislative Study 92. Rome: Food and Agricultural Organisation.

Howarth, W. and McGillivray, D. (2001) *Water pollution and water quality law.* Crayford: Shaw & Sons.

ILA (1986) *The Seoul Rules on international groundwater.* ILA sixty-second conference (Seoul). London: International Law Association.

ILA (2004). *The Berlin Rules on water resources.* ILA seventy-first conference (Berlin). London: International Law Association.

ILC (2008) *Draft articles on the Law of Transboundary Aquifers,* Official Records of the UN General Assembly, Sixty-third Session, Supplement No. 10 (A/63/10). New York: United Nations.

McCaffrey, S.C. (2005) The human right to water. In E.D. Weiss, L.B. de Chazournes and N. Bernasconi-Osterwalder (eds) *Fresh water and international economic law,* International Economic Law Series. Oxford: Oxford University Press.

McCaffrey, S.C. (2011) The International Law's Commission's flawed Draft Articles on the Law of Transboundary Aquifers: The way forward. In G. Eckstein (ed.) Special Issue: Strengthening cooperation on transboundary groundwater resources. *Water International,* 36(5): 566–72.

Mechlem, K. (2003) International Groundwater Law: Towards Closing the Gaps? *Yearbook of International Environmental Law,* vol. 14. Oxford: Oxford University Press.

Meinzen-Dick, R., Pradhan, R., Palanisami, K., Dixit, A. and Athukorala, K. (2004) *Livelihood consequences of transferring water out of agriculture: Synthesis of findings from south Asia.* Washington, DC: International Policy Research Institute.

Misstear, B.D.R., Ashley, R.P. and Lawrence, A.R. (1998) Groundwater pollution by chlorinated solvents: the landmark Cambridge Water Company case. In J. Mather, D. Banks, S. Dumpleton and M. Fermor (eds) *Groundwater contaminants and their migration.* Special Publication 128. London: Geological Society.

Musgrave, W. (2011) Historical development of water resources in Australia: Irrigation policy in the Murray–Darling Basin, In L. Crase (ed.) *Water policy in Australia: The impact of change and uncertainty.* Washington, DC: RFF Press.

Quevauviller, P. (2007) Integrated management principles for groundwater in the WFD context. In P. Quevauviller (ed.) *Groundwater science and policy: An international review.* Cambridge: RSC Publishing.

Reid, E. (2004) The doctrine of abuse of rights: Perspective from a mixed jurisdiction. *Electronic Journal of Comparative Law*. 8(3) 1–15.

Salman, M.A. and Bradlow D.D. (2006) *Regulatory framework for water resources management: A comparative study*. Law, Justice, and Development Series. Washington, DC: World Bank.

Siddiqui, I.A. (1992) History of water laws in India. In C. Singh (ed.) *Water law in India*. New Delhi: Indian Law Institute.

Tarlock, A.D., Corbridge, J.N. and Getches, D.H. (2002) *Water resources management. A casebook in law and public policy*. 5th edn. University Casebook Series. New York: Foundation Press.

Teclaff, L.A. and Utton, A.E. (eds) (1981) *International groundwater law*. London: Oceana Publishers.

Templeman, L. (1994) Water rights diluted: Cambridge Water Company Eastern Leather PLC (Apellants). *Journal of Environmental Law* 6(1): 137–56.

UN (1997) Convention on the Law of the Non-Navigational Uses of International Watercourses. 36 ILM710 (1997). New York: United Nations.

UNCESC (2003) *Substantive issues arising in the implementation of the International Covenant on Economic, Social and Cultural Rights. General Comment No. 15: The right to water (arts. 11 and 12 of the International Covenant on Economic, Social and Cultural Rights)* Committee on Economic, Social, and Cultural Rights, Twenty-ninth session Geneva, 11–29 November 2002, Agenda item 3 E/C.12/2002/1120 January 2003.

UNGA (2008) *The law of transboundary aquifers*, UN General Assembly, 63th session, A/RES/63/124. United Nations, New York. http://internationalwaterlaw.org/documents/intldocs/UNGA_Resolution_on_Law_of Transboundary_Aquifers.pdf

UNGA (2011) *The law of transboundary aquifers*, UN General Assembly, 66th session, 3 October to 11 November 2011, Draft resolution A/C. 6/66/L.24. United Nations, New York. http://www.un.org/en/ga/sixth/66/TransAquifer.html.

Wiber, M.G. (1992) Levels of property rights and levels of law: A case study from northern Philippines, *Man* (NS) 26: 469–92.

Wittfogel, K.A. (1957) *Oriental despotism: A comparative study of total power.* New Haven, CT: Yale University Press.

Yamada, C. (2007) *Third report on shared natural resources: Transboundary groundwaters*. Fifty-Seventh Session, Geneva, 2 May–3 June and 4 July–5 August 2005. UN Doc. A/CN.4/551. Geneva: United Nations General Assembly, International Law Commission.

Yamada, C. (2008) *Fifth report on shared natural resources: Transboundary groundwaters*, Sixtieth session, Geneva, 5 May–6 June and 7 July–August 2008. UN Doc. A/CN.4/591. Geneva: United Nations General Assembly, International Law Commission.

4 An introduction to the economic dimension of groundwater

Economics comprises the third dimension of importance to groundwater, the other two, the physical and the legal dimensions having been described in the preceding chapters. This chapter begins with a description of the concept of 'good'. It continues with a note on the nature of economics and its relation to scarcity and a discussion on the various types of good (economic good, public or merit good, commodities and private goods) and their relationship to groundwater. A section on the economic value of groundwater follows. The chapter closes with a discussion on managing groundwater as a private good.

Groundwater and the concept of good

Aristotle, in his *Politics* (350 BC), considered that the acquiring of possessions (goods) is part of economics (or *oikonomia* meaning household management, from the Greek *oikos* house and *nemein* to manage) 'because without the necessary [goods] it is impossible to live and live well'. The making of money (*chrematistiki*) he did not consider to be the task of economics, as he defined it. Thus, Aristotle related economics to the wellbeing of man, and in this respect water may be considered as a necessary good for without it there is no wellbeing or indeed survival.

Fifteen hundred years later, the Austrian economist Menger, in his *Principles of Economics* (1851) investigated how a thing becomes a good or acquires the character of a good. He suggested four requirements:

1 a human need;
2 such properties as to render the thing capable of being brought into a causal connection with the satisfaction of the human need;
3 human knowledge of the causal connection; and
4 command of the thing sufficient to direct it to the satisfaction of the need.

In general, the idea that goods are those things that are suited to the satisfaction of human needs persists to the present day.

Water, whether on the surface or underground, eminently qualifies as a good. In a broad sense, the human need for groundwater includes not only the usual range of services (drinking and domestic, irrigation, industrial), but also the preservation of the ecological environment (rivers, forests, wetlands) and non-human life. Humans have always been able to use and control groundwater from springs and shallow wells. With modern methods and machinery, strong groundwater flows from artesian aquifers have been harnessed, whilst deep groundwater, which was inaccessible in the past, has now become available and put to various uses for the benefit of man. Groundwater that becomes polluted loses its property to satisfy human need as a source of potable supply. However, with modern technology, it can become wholesome again and fit for human consumption after treatment, although at a cost. Flowing water in rivers was, at one time, the main source of energy driving the corn and textile mills of England and the eastern USA. This need no longer exists, and in this sense, flowing water has ceased to be a good. However, flowing water has now become a source of hydroelectric energy, and therefore a good. Steam is no longer needed to power railway engines (although, many power plants still run on steam turbines) nor is ice (harvested from natural resources) needed to provide cooling or refrigeration, and therefore, both have lost their character as goods.

Economics and water scarcity

Economics according to Robbins (1932) is 'the science which studies human behaviour as a relationship between ends and scarce means which have alternative uses'. Thus, economics is concerned with scarcity, i.e. the condition in which the requirements for a good are greater than available quantities. This is also the modern understanding (McEachern, 2006). It differs from Aristotle's idea of *oikonomia* (he does not seem to have thought of it as a science), although the human element to economics is still very much evident.

Water scarcity is a relatively recent phenomenon that has arisen primarily in the Indian subcontinent, China and the western USA due to the overexploitation of groundwater mainly for irrigation (see Chapter 2). The Dublin Statement on Water and Sustainable Development of the International Conference on Water and the Environment (ICWE, 1992) recognised the economic dimension of water in its fourth principle: 'Water has an economic value in all its competing uses and should be recognised as an economic good'. And in a supplementary note:

> Within this principle [meaning principle no 4], it is vital to recognise first the basic right of all human beings to have access to clean water and sanitation at an affordable price. Past failure to recognise the economic value of water has led to wasteful and environmentally damaging uses of the resource. Managing water as an economic good is an important way of achieving efficient and equitable use, and of encouraging conservation and protection of water resources.

The principle has attracted controversy as opening the way to treating water as a commodity. Nevertheless, the intention seems to have been that the recognition of an economic dimension to water can promote its more sustainable and efficient use.

The concept of economic good and groundwater

Scarcity in relation to man's wants is the condition that defines an economic good. A good becomes an economic good when the requirements for it over a period are larger than the available quantity. Conversely, a good becomes a non-economic good when the requirements for it are smaller than the available quantity.

Perry *et al.* (1997) suggested that the idea that water is an economic good follows from Robbins's definition of economics for two reasons: first, as water serves a multiplicity of ends, ranging from drinking and bathing, through irrigation, recreation, and environmental use, to waste, it satisfies the condition of 'alternative uses'; second, as water may at times be considered scarce in the sense that it cannot fully satisfy all its alternative uses simultaneously, it satisfies the condition of scarcity. McNeill (1998), like Perry *et al.*, considered water an economic good because it is a scarce resource for which there are many competing uses. Savenije and van der Zaag (2002) suggested that water, owing to a combination of characteristics including its importance to human life and the environment, is not a normal but a special economic good. This is reminiscent of Menger's use of drinking water to illustrate that some economic goods because of their importance are provided by government in such large quantities that they lose their economic character, occupying an intermediate position between economic and non-economic goods. Quoting from Menger,

> pure healthy drinking water [which is] considered ... of such importance by the inhabitants of many cities that, wherever nature does not make it abundantly available, it is brought by aqueducts to the public fountains in such large quantities that not only are the requirements of the inhabitants for drinking water completely met but also, as a rule, considerable quantities above these requirements are available.

This would seem to imply that water as a good that is essential to humans for their survival transcends the concept of an economic good and also that governments should avoid scarcity by ensuring that it is made available to all members of society, even the poorest.

Groundwater, even when rechargeable, is increasingly becoming an economic good. As already discussed in Chapter 2, in recent years groundwater use in many countries, mainly for irrigation, has exceeded available quantities, and is no longer able to satisfy all requirements without permanent depletion. For rechargeable aquifers with sufficient storage capacity to buffer periods of drought, available quantities may be taken to be the equivalent of the average recharge of one or more years. In thin aquifers with little storage, available quantities are much less and usually limited to annual or seasonal recharge. Fossil aquifers receive no recharge and thus available groundwater quantities are dependent solely on storage in a manner similar to ores or hydrocarbons. Thus, although they may be able to satisfy requirements, they are only capable of doing so over finite periods with depletion setting in at the outset. It is prudent, therefore, to consider fossil groundwater as an economic good, especially when its exploitation is on a large scale.

Groundwater in its different economic guises

Groundwater as a public good, private good or an intermediate good

Goods that human beings use to satisfy their needs usually fall into a particular category by virtue of their characteristics. So, air in the atmosphere, for example, is generally considered to be a public good, as its consumption neither exhausts its supply nor is it easy to stop others from consuming it. It is also free. In economics jargon, a public good is neither rival nor exclusive. Rainfall, for as long as it lasts, falls into this category.

Diametrically opposite to public goods are private goods. Again, using economics jargon, private goods are rival and exclusive. A manufacturer or supplier of private goods can easily exclude people who do not pay from using them. Also, the quantity available for consumption is limited (unless the supplier floods the market), which means that one unit consumed by one person leaves one unit less for the next person. Mineral bottled and similar aerated waters which have been captured usually from groundwater sources (springs or boreholes) are private goods because, first, the owner or supplier of the mineral water can easily exclude non-paying customers, and second, because the consumption of a bottle of mineral water leaves less for others to consume.

Groundwater does not fall easily into either category. It cannot be a public good resource because, even when renewable, it is finite. Fossil groundwater, when abstracted, is not replaced, leaving others with less to consume. Renewable groundwater becomes depleted when extraction exceeds recharge, but it too cannot be a private good. This is because as a common pool resource, it is difficult to exclude others from using it. For example, when a person takes water from his well or borehole, he also, to a smaller or a larger degree, takes a portion of his neighbour's water. There is no easy physical way of stopping this, and as discussed in Chapter 2, when many users tap the same aquifer, it often leads to overexploitation and depletion. This is a well-known phenomenon in other similar common pool natural resources, such as the fish in the oceans or large forests. Problems of common pool resources were highlighted by Hardin (1968). Since then there have been numerous publications (McCay and Acheson, 1987; Ostrom, 1990, 2008; The National Research Council (NRC), 2002, and specific to groundwater, Brentwood and Robar, 2004).

Thus, groundwater belongs to an intermediate category of goods, which is neither public nor private, known as open access goods. As their utilisation cannot be constrained by private ownership to prevent overuse and adverse impacts on the environment, they are usually regulated by government. As discussed in Chapter 3, in the case of groundwater, this is generally achieved by the issue of abstraction permits.

Groundwater as a merit good

The third main category of goods is merit goods, a concept first developed by Musgrave (1956, 1959), and much debated and added to since (see Van Eecke, 2006, for an anthology of articles and commentary). The significance of merit goods is not because of their economic dimension – although, if one is to follow Aristotle they are part of economics by virtue of the fact that they are necessary for 'the living and well living' of men – but because of an underlying ethical dimension which requires public bodies to provide such goods for the benefit of society at large, even if individuals themselves are not able to evaluate such wants correctly. There is some similarity between Menger's idea of intermediate economic goods and Musgrave's concept of merit goods, which as mentioned above, Menger illustrated by reference to drinking water. There is no doubt that groundwater, as a source of water for drinking, sanitation and other human uses, is a merit good, even after it is captured and processed by private utilities. Governments ensure that individuals, even when they are not able to pay, cannot be denied access to it.

Groundwater as a commodity or product

The term commodity has different shades of meaning. However, most people identify commodities as goods that are traded on the market. Other characteristics are that they are standardised, traded in bulk and have units that are interchangeable (Black, 2002). Examples of commodities are oil, metals and agricultural products, all of which are traded on local or international markets. Although groundwater, and for that matter water, shares characteristics with other commodity goods, it is not traded on any commodity exchange in the same way as, for example, oil.

A product is an article or substance that is manufactured or refined for sale (*Oxford English Dictionary*, 2007). Thus, groundwater becomes a product when it is specifically manufactured or refined for the sole purpose of sale, as is the case with bottled mineral waters, or waters in soft or aerated drinks.

There has been considerable resistance to accepting the idea that water, including groundwater, in its natural state should be treated as a commodity or product. Much of this resistance has been on ethical grounds and relates to the special character of water as a good that is necessary for life that has no substitute. The EU Water Framework Directive (EC, 2000) expressed this clearly as follows: 'Water is not a commercial product like any other, but rather, a heritage which must be protected, defended and treated as such'. Others, while recognising that water in times of scarcity is an economic good, commented that the value of water to the community and the environment overrides the narrow economic considerations of the market and market prices (Brown, 1997; McNeill, 1998).

Whether water in its natural state (meaning as it occurs in rivers, lakes, aquifers etc) is a 'good' or 'product' has been the subject of a continuing debate arising from the North American Free Trade Agreement (NAFTA, 1992) of the USA, Canada and Mexico (Weiss, 2005; Coffin *et al.*, 2011). The debate arose primarily because there is no clear definition of 'good' or 'product' either in the NAFTA or in the General Agreement on Tariffs and Trade (GATT, 1994) to which Article 202 of NAFTA makes reference to for a definition. Water in GATT is mentioned in heading item 22.01 of its Harmonised Commodity Description and Coding System, which reads: 'waters, including natural or artificial waters and aerated waters, not containing added sugar or other sweetening matter or flavouring, ice and snow'. This and the explanatory note that states that the heading item covers 'ordinary natural water of all kinds (other than seawater)' led some people to interpret NAFTA to mean that water in its natural state may be treated as a tradable good or product that can be transferred in bulk across borders. This interpretation was refuted in the 1993 joint statement of the governments of Canada, Mexico and the USA which included the following statement:

> Water in its natural state in lakes, rivers, reservoirs, aquifers, water basins and the like is not a good or product, is not traded, and therefore is not and never has been subject to the terms of any trade agreement.

This position was reaffirmed by the International Joint Commission (IJC, 2000) which, however, pointed out that 'When water is "captured" and enters into commerce, it may, however, attract obligations under GATT, FTA, and NAFTA'.

The economic value of groundwater

A comprehensive discussion of groundwater valuation has been presented by the National Research Council (NRC, 1997). Earlier, Custodio and Gurgui (1989) discussed groundwater economics and more recently Koundouri *et al.* (2003) and Job (2006).

Classical economists are fond of using water to illustrate their economic ideas. Menger has already been mentioned with regard to using water as an example of goods changing their economic character. About 100 years earlier, Adam Smith (1776) used the paradox of water and diamonds to explain the concept of value. In his own words:

> ... value ... has two different meanings, and sometimes expresses the utility of some particular object, and sometimes the power of purchasing other goods which the possession of that object conveys. The one may be called 'value in use'; the other, 'value in exchange'. ... Nothing is more useful than water; but it will purchase scarce anything; scarce anything can be had in exchange for it. A diamond, on the contrary, has scarce any value in use; but a very great quantity of other goods may frequently be had in exchange for it.

Diamonds fetch a high price because they are scarce; water has a low price because it is abundant. But water has a high value because it is necessary to existence. Diamonds have a low value because they are hardly necessary to existence. In modern economic terms, water has a higher total utility (this is the total satisfaction derived from consumption) than diamonds. However, its marginal utility (this is the increase or decrease in satisfaction from consuming an additional unit of a good) is lower. Simply expressed, most people can quench most of their thirst after the first glass of water, and so the utility of the first glass is very high. But this soon decreases to zero after the third or fourth glass. In contrast, people continue to value their second and third and fourth diamonds as much as their first. After all, as the saying goes, 'diamonds are a girl's best friend'.

The concept of total economic value

Until relatively recently economic valuations focused on marketed goods. Valuations of natural resources were generally neglected. In the early 1980s, the concept of total economic value (TEV) was developed, primarily in response to damage inflicted by humans on the environment. There have been somewhat different views expressed on the taxonomy of TEV (Randall and Stoll, 1983; Fisher and Raucher, 1984; Boyle and Bishop 1987; Freeman, 1993/2003; Plottu and Plottu, 2007), but generally the TEV of a resource is considered to be the sum of two components, a use value and a non-use value. The use value comprises the direct or indirect use of a resource. The non-use value is less tangible. It was first introduced by Krutilla (1967) and reflects people's sense of wellbeing derived from the knowledge that a natural resource exists (existence value or intrinsic value or passive value) and their desire to bequeath the natural resource to future generations (bequest value) or to preserve the option for the future use of a resource (option value). This third value, first introduced by Weisbrod (1964), is usually included as part of the use component on the basis that it involves the utilisation of the resource, even if this were to happen in the distant future.

The concept of TEV was used by the National Research Council (NRC) as a basis to value groundwater (NRC, 1997). In the NRC groundwater valuation terminology the TEV taxonomy of use and non-values is replaced by a 'physical state terminology' of extractive values and in-situ values, possibly in the light of comments by Freeman (1993) on the inconclusiveness of use and non-use definitions. The extractive values are all direct-use values and include the costs associated with the municipal, industrial, agricultural and other extractive services provided by groundwater. The in-situ values include indirect use values (items a, b and c) and non use values (item d), as follows:

a   ecological and recreational values; these relate to groundwater flows that help to maintain wetlands and rivers, and their ecology;
b   buffer value; groundwater may be used to alleviate periodic shortages in surface water supplies;
c   land subsidence and sea water intrusion protection values; these relate to the excessive abstraction of groundwater in confined aquifers that might cause land subsidence and in coastal aquifers contamination with sea water; and,
d   existence and bequest values, described above.

Other considerations in the valuation of groundwater may include the impact of diminishing groundwater reserves and of deepening water tables to domestic users, mainly in third-world countries, who rely for their water supply on shallow dug wells. Also, of course, the TEV should reflect the importance of water in maintaining life on the planet.

Groundwater, due to its purity, natural storage and, usually, proximity to demand, is generally cheaper than treated surface water and invariably, much cheaper than desalinated sea water. Unless subsidised, the price that consumers pay reflects primarily its extraction value, plus profit to shareholders in the case of private undertakers. One of the ways in which governments try to include in-situ values is by applying a fee to water undertakers in the form of an abstraction licence, which varies with the type of use. Recently, the Environment Agency of England and Wales introduced an environmental improvement charge (EA, 2011/2012). There is generally no increase in price to the consumer to reflect scarcity during periods of drought, the preferred solution being to reduce supply or prohibit certain types of uses (washing cars, irrigating lawns) or both. However, use of water during the summer can attract a higher licence fee. In extreme cases, (see the Edwards aquifer in Texas in Chapter 5), in order to maintain ecological integrity authorities purchase the land where groundwater recharge occurs or pay abstractors to reduce pumping.

Determination of economic value

For marketed goods or commodities, price is normally more or less equivalent to economic value. Thus, the price of oil or metals on the international market reflects their economic value. Groundwater is not a marketed good in the same sense, and therefore its economic value has to be determined by other means. In the rare cases where a groundwater resource has no in-situ value, the estimation of its economic value is relatively straightforward as it relates primarily to production costs and, when more than one sector is involved, opportunity costs. Opportunity cost is the value of the best forgone alternative. This means that the cost of a good is not just its unit cost but all of the value of what has been given up in the course of acquiring the good. In the case of groundwater, opportunity cost is the value of groundwater in its highest value alternative use. For urban water supplies opportunity costs are low, and extraction costs are high. In contrast, good-quality irrigation groundwater has high opportunity costs and, usually, low extraction costs. The price to consumers of urban water supply in industrial countries more or less reflects the economic cost of water. However, the price of irrigation water is usually significantly lower, not only because farmers are often subsidised, but also because opportunity costs are not taken into account. Based on this premise, irrigation water is undervalued in comparison to its economic cost. Gibbons (1986) carried out a comprehensive study of the economic value of water which focused on extractive values for different sectors in the USA, including municipal, irrigation and industry. Her study indicated that water values varied from sector to sector and also within the sectors themselves. In general, irrigation water had the lowest value, but also the greatest variability depending on crop type. She also found that municipal water demand was inelastic (not responsive) to price changes (the average price elasticity for municipal water demand in the eastern USA was –0.37 and for the arid western USA –0.54). Demand for irrigation water also appeared to be price-inelastic and likely to remain so until such a time as water costs rose dramatically. Where irrigation water costs have been kept artificially low through direct subsidies or cheap energy, economic rents (these are earnings greater than the cost to keep the resource in its present use) have been high. However, she noted that, as water costs rise, these rents will be reduced and farmers will have a greater incentive to conserve water or to consider alternate crops.

Another parameter to consider in groundwater pricing is 'scarcity rent'. It is a type of economic rent that reflects future costs or the cost to future generations of using up a finite resource to-day. Pumping groundwater in excess of recharge sooner or later leads to the depletion of its stock. In such conditions, groundwater earns economic rent by virtue of its scarcity. An economically efficient price should, therefore, include for both the marginal cost of extraction and scarcity rent. Scarcity rents should be considered in the pricing of fossil groundwaters, although in very large aquifers depletion may take a long time to develop. Scarcity rents are difficult to estimate, but ignoring them encourages over-exploitation and wastage. A model to estimate scarcity rent was developed by Moncur and Pollock (1988), who used the groundwater abstraction from the Honolulu aquifer to illustrate their result. Koundouri (2003) compared two different methodologies that allowed the indirect estimation of shadow scarcity rents. She used the Kiti aquifer, Cyprus, to compare results.

Difficulties in the determination of economic value arise when in-situ values play an important role, for which a monetary value cannot be easily calculated. A number of methods have been used, all seeking to establish the monetary value that users attach to changes in the services provided by a resource. They are all based on the principle of willingness to pay for a good (WTP) or willingness to accept compensation for giving it up (WTA). Theoretically, the two should be identical. In practice, there is often a difference because individuals would tend to put their own interest first, receiving payment being usually the more attractive option. WTP represents the equilibrium market price at the margin of potential buyers of the good or service, i.e. the individual's best offer for purchasing the increment. Methods used to determine TEV include: cost of illness, travel cost, averting behaviour, hedonic pricing and contingent valuation. None of these provides a satisfactory answer to determining the TEV of groundwater, and only one, the contingent valuation method (CVM), seems to provide (controversially) an indication of non-use values (NRC, 1997), although estimates are difficult to validate externally. The CVM method is based on answers to questions by respondents on the monetary values they attach to hypothetical changes in the service provided by a good. Answers from respondents may not always be reliable, and may be biased in favour of the individual's own interest. Thus questions must be carefully phrased and the various aspects related to the resource and its benefits as may apply to a particular situation carefully and objectively described. A similar method which has been used to elicit TEV is the choice experiment method (CE) (Bateman *et al.*, 2002). It differs from CVM in that respondents are asked to make choices between hypothetical alternatives. CE derived values tend to be much greater than those by CVM (Boyle *et al.*, 2004). For example, in Denmark, Hasler *et al.* (2005) undertook CVM and CE studies to determine willingness to pay to maintain the purity of groundwater and, also very good groundwater conditions for plant and animal life. The CE value was US$517 (Danish kroner (DKK) 3,103) or more than four times the CEV (Table 4.1).

There are only a few examples of groundwater valuations, none of them comprehensive, and most addressing a particular aspect, mainly related to contamination and water quality. NRC (1997) in their review of previous studies in the USA referred to five studies in which the averting behaviour method was used and 10 that used the CVM method. All studies addressed groundwater contamination. There was a wide range of avoidance costs from about US$120 to more than US$1,000 per year per household. The CVM studies also indicated a wide range of values of between approximately US$14 to near US$1,000 per year that people were willing to pay to reduce contamination. A variation in values has also been found in studies in Europe and elsewhere (Table 4.1).

In an interesting study in the Madachi farming area of the Hadejia-Jama river basin, Northern Nigeria, Acharya and Barbier (2000) undertook an economic analysis of the impact of falling groundwater levels on irrigated dry season vegetable and wheat production. They estimated that the welfare impact on dry season farmers due to a one metre fall in groundwater level was a loss of US$32.50 per year per farmer for vegetable farmers and US$331 per year per farmer for wheat and vegetable farmers. They estimated the welfare loss per household to be US$0.12 per day (US$44 per year). The estimated annual value of maintaining groundwater levels throughout the basin was US$4.76 million, which provided an indication of the value for maintaining groundwater recharge from the floodplain wetlands.

Birol *et al.* (2010) investigated the economic benefit of a plan to recharge with tertiary-treated sewage effluent the Akrotiri aquifer, west of the town of Limassol, Cyprus. For many years this aquifer has been used for irrigation of citrus, vines and vegetables. Overabstraction has led to a decline in groundwater levels, sea water intrusion and a negative impact on the hydro-ecology of the nearby Akrotiri salt lake. The CE survey indicated a willingness to pay by farmers of on average US$0.034 (CYP (Cyprus pounds) 0.014, using the exchange rate of 2 September, 2009) per cubic metre of water per household to maintain groundwater quality, US$0.068 (CYP0.028) per cubic metre to maintain groundwater quantity and US$0.00049 (CYP0.0002) per cubic metre to maintain an extra job in agriculture. In contrast, Limassol residents were willing to pay significantly more than farmers: US$0.32 (CYP0.133) per cubic metre to maintain groundwater quality, US$0.00098 (CYP0.0004) per cubic metre to maintain an extra job in agriculture and US$0.13 (CYP0.055) per cubic metre to maintain ecological conditions. The low number of protestors (six per cent of the sample) suggested that farmers accepted the use of treated effluent for artificial recharge, although Limassol residents were likely to select alternatives to maintain the current groundwater

Table 4.1 Summary of groundwater valuations using CVM

Country	Description	Willingness to pay (WTP) per year per household		Author
		US$	Local currency	
USA	Various aquifers	14–1000		Various authors; summarised in NRC (1997)
Europe	Italy: economic value of groundwater quality of aquifer underlying the Milan area	417 (in 1996 dollars)	645,000 Italian lire	Press and Söderqvist (1998)
	France: preservation of groundwater quality of Alsatian aquifer	104 (in 1998 dollars)	617 French francs	Stenger-Letheux and Willinger (1998)
	France: restoration of groundwater quality of Upper Rhine aquifer	53–95[a] (in 2006 dollars)	42–76[a] euros	Aulong et al., (2006)
	Denmark: maintaining groundwater quality of aquifers	119[b] (in 2005 dollars)	711 Danish kroner	Hasler et al. (2005)
	Spain: economic value of Gavilán aquifer which is used by farmers and supports important wetland	0.63 per m^3 of water (in 2011 dollars)	0.454 euros per m^3 of water	Martinez-Paz and Perni (2011)
Lebanon	Byblos district: farmers WTP to improve the quality of groundwater of the coastal aquifer	100–170		Chami et al. (2008)
Philippines	Demoy aquifer; prevent deterioration of quantity and quality	17–20[c] (in 2007 dollars)	799–929[c] pesos	Martinez and Prantilla, (2007)

Notes

a   53 US$ (42 euros) to restore groundwater quality to drinking water standards and 95 US$ (76 euros) to restore natural quality. (Method: linear logic model)

b   includes maintaining the purity of groundwater and very good groundwater conditions for plant and animal life; the CE was US$ 518 (DKK 3,103) comprising US$ 317 (DKK 1,899) and US$ 201 (DKK 1,204) for maintaining very good conditions for plant and animal life.

c   US$ 17 (799 pesos) per year by domestic users and US$ 20 (929 pesos) per year by commercial users and institutions quality. The authors noted that respondents were probably biased by the severe drought of recent years. The survey does not seem to have addressed the question of the potential risk of biological pollution of the aquifer.

Finally, Liu *et al.* (2009) used CE to determine farmers' preferences in addressing groundwater level decline due to over-exploitation of the Merguellil river basin aquifer in Tunisia. The survey found that most farmers were willing to accept management policy changes to protect the quantity and quality of the resource. However, the better-off farmers, who seem to benefit most from the current regime of over-exploitation, were likely to oppose changes.

The discussion above has highlighted the significance of the environmental and non-use values in the economic valuation of groundwater. But it has also shown that establishing monetary values based on market methodologies of WTP and WTA may be flawed, as these often reflect the individual's preferences. Such preferences may be biased by self-interest, which may not always coincide with the collective benefit to society or non-human entities. Also, the non-visible nature of groundwater may have an effect on its valuation by respondents (Becker and Tonin, 2001). Nevertheless, economic valuations and monetary values are a useful tool that provides information on how scarce resources may be efficiently managed.

Managing groundwater as a private good and externalities

The management of groundwater as a private good arises from economic efficiency considerations. Where there are competing uses, economic efficiency dictates that allocation should go to the best price option. Thus, economic value increases, and in an ideal market of many sellers and many buyers, equilibrium of supply and demand is achieved. The large volumes of cheap groundwater used in irrigated agriculture prompted some economists to propose a market approach to the management of groundwater resources (Winpenny, 1994; Briscoe, 1996). The advantages were considered to be a more sustainable use of groundwater resources, a reduction in wastage, and the option of moving groundwater out of irrigated agriculture to public supply or industry where it would attract a higher economic value.

The market approach to groundwater management presumes that groundwater should be treated as a private good. But, as already discussed, groundwater is not a private good. It has the attributes of an open access good and a merit good, and as such it is susceptible to externalities. Private goods do not admit to externalities: any costs or benefits that fall on a third party are ignored by the two parties to the market transaction. Industry has always had an economic incentive to ignore any pollution to surface water or groundwater caused by the discharge of untreated effluents. As a result, it has often been the case that government has had to step in and bear the cost of damage to the environment or the resource.

There may be circumstances under which the market is able to deal with externalities without the need for government intervention. Coase (1960), in what became known as the 'Coase theorem', suggested that rational individuals may be able to deal with externalities themselves by negotiating an economically efficient solution. There are three conditions for a successful outcome:

1   the number of participants must be small;
2   bargaining costs are very low or zero;
3   property rights to the good are well defined, though irrespective of which party is assigned the property right.

None of these conditions easily fits into groundwater. A very recent study in Denmark (Abildtrup *et al.*, 2012) involving the negotiation between farmers and Danish waterworks concluded that the application of the Coase theorem where these conditions were not met did not result in economic efficiency in the protection of groundwater quality. Government intervention was considered to be the more cost-effective and efficient approach.

The greatest impediment to the Coase theorem in its application to groundwater is, however, the multiplicity of abstractors or the large number of parties that may be affected by the externality. As already discussed in Chapter 2, in India and China, there are hundreds of thousands of abstractors exploiting the same aquifer, which makes it very difficult for parties to negotiate the externalities. Evidence lies in the fact that groundwater levels continue to decline and storage reserves continue to be depleted in many of the world's aquifers. Bargaining costs, especially when lawyers have to be involved, are not usually low and seldom zero. Finally, groundwater rights in many countries are not well defined. Even when they are well defined, questions of equity may arise where poor communities or individuals are not able to remove or reduce the externality by compensating the person or entity responsible. The Coase theorem is concerned with situations of 'now' that affect a small number of human beings directly. In such situations, it may be able to provide an economic solution in the form of a monetary compensation to the injured party, but without necessarily removing the externality. It is difficult to envisage how impacts on groundwater non-use values or ecological impacts may be addressed, as these are not normally of direct concern to individuals but more to society in general.

Gisser and Sanchez (1980) investigated welfare loss due to deepening groundwater levels arising from farmers pumping from the same aquifer. They developed two mathematical models: the first representing 'no control on pumping', in which farmers could pump as much groundwater as they liked. In this model, groundwater allocation follows the absolute ownership doctrine (see Chapter 3; Gisser and Sanchez described it as 'free market behaviour'). The second was based upon 'optimal control', in which the temporal allocation of groundwater is managed. This allows for reserves to be stored for future use, which enables farmers over time to maximise the present values of all their future income streams derived from irrigation. The aquifer was idealised as having a very large storage capacity in relation to recharge, a steady state flow condition (i.e. equilibrium groundwater levels), and fixed inputs of recharge and irrigation returns. Until this work was undertaken, the conventional thinking had been that the temporal allocation of groundwater in a free market where all farmers could pump as much water as they liked from a common aquifer would lead to welfare losses. However, comparison of the two models suggested the opposite. Providing an aquifer had a large enough storage capacity, Gisser and Sanchez found that there was no significant difference in welfare loss between the two management regimes. The authors tested their model on the Peco basin aquifer, New Mexico, and predicted that 'optimal control of groundwater would not enhance the welfare of farmers compared with a strategy of free markets'. In both cases the derived value of future income streams was approximately the same, US$3.1 million. There are two main hyrologeological limitations to the model, both of which were identified by the authors. The first, relates to the requirement for a very large storage or a bottomless aquifer, which allows for steady state equilibrium to be reached. In fact, most aquifers are finite in extent and groundwater levels continue to decline with pumping. The second limitation arises from the fact that recharge is generally variable and not fixed, as assumed in the model.

Koundouri (2004) in a comprehensive review of previous empirical studies was able to confirm that, in general, the Gisser–Sanchez model prediction of negligible benefits from optimal groundwater management held true. Benefits from management were found to be moderately sensitive to aquifer storage capacity and particularly sensitive to the financial factors of interest rates and the demand over price ratio. However, the model addresses only the specific externality of welfare loss by farmers due to deepening groundwater levels. A free-for-all pumping regime can cause many other externalities: groundwater depletion and quality deterioration, the loss of the buffer function of groundwater during periods of drought or of its potential use by future generations, sea water intrusion and ground subsidence. Some of these may be irreversible. A management strategy can help address such externalities and result in positive benefits. For example, future decreases in discount rates would favour a management approach that preserves groundwater so that it remains available for use by future generations.

Concluding remarks

In the past there was ample groundwater to satisfy all needs. In the last 60 years or so mechanisation has brought about an explosive increase in its exploitation, especially in the agricultural sector. The need for food acted as a catalyst and the use of groundwater, a convenient and cheap source of good-quality water, was generally encouraged by governments, often through energy subsidies. However, the impacts, which went largely ignored, were significant: deepening groundwater levels, depletion of resources, contamination of coastal aquifers, land subsidence and reductions of flows to rivers and wetlands. The consequences of cheap or undervalued groundwater kindled interest in its economic nature, and particularly whether or not it could be managed on economic principles, including its trading in the market place.

Groundwater is not easy to classify in economic terms. It is probably best thought of as an open access good: as a common pool resource others cannot be easily excluded from using it, and as a finite resource, even when renewable, its consumption leaves less for others to enjoy. Groundwater has also the characteristics of a merit good, and governments have a responsibility to ensure that people, even when they are not able to pay, cannot be denied access to it. This may also be extended to the natural world, whose survival is so vital for the well being of human beings and of life itself.

In its natural state, groundwater is generally not considered to be a commodity or product. This is because of the special character of water as a good that is necessary for life that has no substitute. Groundwater becomes a commodity or product after it is captured and processed for the sole purpose of sale, as is the case with bottled mineral waters or waters in soft or aerated drinks.

That groundwater may have an economic relevance is related to scarcity, which is the condition in which requirements by users exceed available quantities. With scarcity, groundwater takes the character of an economic good and has an economic value. The total economic value (TEV) of groundwater comprises extractive values and in-situ values. The former relate to its direct use for municipal, agricultural or industrial purposes. In-situ values relate mainly to its indirect use by the hydro-environment, as this relates to man. Non-use values, although important, are a little more nebulous and relate to the enjoyment of groundwater by future generations. There is also the philosophical question of whether economic valuations should take account of groundwater uses by nature outside man. Establishing extractive values is relatively easy as these are generally based on tangible costs. In-situ values are more difficult to determine and approaches based on market methodologies of willingness to pay (WTP) and willingness to accept (WTA) compensation have been developed. The contingent valuation method (CVM) is said to capture the TEV, although this is controversial, especially when in-situ values play an important role. There are only a few examples of TEV for groundwater and almost all relate to WTP with regard to contamination. Derived monetary values using WTP and WTA often reflect the individual's preferences, which may be biased by self interest that may not always coincide with the collective benefit to society or non-human entities. Nevertheless, economic valuations and monetary values constitute a useful tool that provides information on how scarce resources may be efficiently managed.

Groundwater has competing uses. In an ideal situation of many sellers and many buyers, a market approach should bring an economically efficient allocation to groundwater and an increase in its economic value. In practice, this means that subsidised cheap groundwater for irrigation becomes reallocated to the higher priced municipal or industrial supplies. Private goods respond well to market forces, but there are reservations for goods such as groundwater which are susceptible to negative externalities. The market is not very good at responding to externalities and almost invariably adverse impacts require government intervention. Economic efficiency is but one aspect in the economic dimension of groundwater. Other aspects which are as important, if not more, are the benefits that groundwater brings to all humans, rich and poor, the environment and non-human life. Allowing groundwater to be traded by making groundwater rights transferable presents a more complex set of issues and challenges than a simple market transaction. This is the subject of the next chapter.

References

Abildtrup, J., Jensen, F. and Dubgaard, A. (2012) Does the Coase theorem hold in real markets? An application to the negotiations between waterworks and farmers in Denmark, *Journal of Environmental Management*, 93(1): 169–76.

Acharya, G. and Barbier, E. (2000) Valuing groundwater recharge through agricultural production in the Hadejia-Nguru wetlands in northern Nigeria, *Agricultural Economics*, 22: 247–59.

Aristotle (2007 [350 BC]) *Politics*, translated by B. Jowett. Adelaide: The University of Australia.

Aulong S., Rinaudo, J.D. and Bouscasse, H. (2006) *Assessing the costs and benefits of groundwater quality improvement in the upper Rhine valley quaternary aquifer (France)*. Deliverable D25, Rapport BRGM/RP-55061-FR. Orléans: Bureau de Recherches Géologiques et Minières. Available at www.wfd-bridge.net

Bateman, I.J., Carson, R.T., Day, B., Hanemann, M., Hanley, N. *et al.* (2002) *Economic valuation with stated preference techniques: A manual*. Cheltenham: Edward Elgar.

Becker, N. and Tonin, S. (2001) Issues in the valuation of groundwater benefits. In E. Dosi, (ed.) *Agricultural use of groundwater*. The Hague: Kluwer Academic Publishers.

Birol, E., Koundouri, P. and Kountouris, Y. (2010) Assessing the economic viability of alternative water resources in water-scarce regions: Combining economic valuation, cost-benefit analysis and discounting, *Ecological Economics*, 69: 839–47.

Black, J. (2002) *Oxford Dictionary of Economics*. Oxford: Oxford University Press.

Boyle, K.J. and Bishop, R.C. (1987) Valuing wildlife in benefit cost analyses: A case study involving endangered species. *Water Resources Research*, 35(5): 943–50.

Boyle, K.J., Morrison, M. and Taylor, L.O. (2004) Why value estimates generated using choice modelling exceed contingent valuation: Further experimental evidence. Paper presented at the Australian Agricultural and Resource Economics Society Conference, Melbourne.

Brentwood, M. and Robar, S.F. (Eds) (2004) *Managing common pool groundwater resources: An international perspective*. Westport, CT: Praeger.

Briscoe, J. (1996) Water as an economic good: The idea and what it means in practice. In *Proceedings of the World Congress of the International Commission on Irrigation and Drainage*, Cairo, Egypt, September.

Brown, F. L. (1997) Water markets and traditional water values: Merging commodity and community perspectives. *Water International* 22(1): 3–4.

Chami, D. El., Moujabbar, M. El. and Scardigno, A. (2008) The contingent valuation method for economic assessment of groundwater: A Lebanese case study, *Mediterranean Journal of Economics, Agriculture and Environment*, VII(3): 19–24.

Coase, R (1960) The problem of social cost, *Journal of Law and Economics*, 3: 1–44.

Coffin, V., Poulton, D.W. and Ploeg, C.V. (2011) *Our water and NAFTA: Implications for the use of market-based instruments for water resources management*, Canada West Foundation Environment and Research Series. Calgary: Canada West Foundation.

Custodio, E. and Gurgui, A. (eds) (1989) *Groundwater economics: Selected papers from a symposium held in Barcelona, Spain*. Amsterdam: Elsevier Science.

EA (2011/2012) *Abstractions charges scheme 2011/2012*. Bristol: Environment Agency.

EC (2000) Establishing a framework for Community action in the field of water policy. Directive 2000/60/EC of the European Parliament and of the Council of 23 October 2000, *Official Journal of the European Union* L327, 22/12/2000 P001–0073.

Fisher, A. and Raucher, R. (1984) Intrinsic benefits of improved water quality: Conceptual and empirical perspectives. In V.K. Smith and D.A. Witte (eds) *Advances in applied microeconomics*. Greenwich, CT: JAI Press.

Freeman, A. M. (2003) *The measurement of environmental resources values: Theory and methods*. 2nd edn. Washington, DC: Resources for the Future.

GATT (1994) General Agreement on Tariffs and Trades, World Trade Organisation (WTO) http://www.wto.org/english/docs_e/legal_e/06-gatt_e.htm

Gibbons, D.A. (1986) *The economic value of water*. Washington, DC: Resources for the Future.

Gisser, M. and Sanchez, D.A. (1980) Competition versus optimal control in groundwater pumping, *Water Resources Research*, 16(4) 638–42.

Hardin, G. (1968) The tragedy of the commons, *Science* (NS) 162(3858): 1243–48.

Hasler, B., Lundhede, T., Martinsen, L., Neye, S. and Schou, J. S. (2005) *Valuation of groundwater protection versus water treatment in Denmark by choice experiments and contingent evaluation*, NERI Technical Report No. 53. Copenhagen: National Environment Research Institute, Ministry of the Environment.

ICWE (1992) *The Dublin Statement on water and sustainable development*. Dublin: ICWE.

IJC (2000) Protection of the waters of the Great Lakes: Final Report to the Governments of Canada and the United States. http://www.ijc.org/boards/cde/finalreport/finalreport.html

Job, C.A. (2006) *Groundwater economics*. New York: Taylor & Francis.

Koundouri, P. (2003) Contrasting different methodologies to deriving natural resources scarcity rents: Some results from Cyprus. In P. Koundouri, P. Pashardes, T. Swanson and A. Xepapadeas (eds) *Economics of water management in developing countries: Problems, principles and policies*. Cheltenham: Edward Elgar.

Koundouri, P. (2004) Current issues in the economics of groundwater resource management, *Journal of Economic Surveys*, 18(5): 703–40.

Koundouri, P., Pashardes, P., Swanson, T. and Xepapadeas, A. (eds) (2003) *Economics of water management in developing countries: Problems, principles and policies*. Cheltenham: Edward Elgar.

Krutilla, J.V. (1967) Conservation reconsidered, *American Economic Review* 57: 787–96.

Liu, X., Mchrrafiyeh, H., Noden, D., Sarr, M. and Swanson, T. (2009) An analysis of water users' preferences for a community based management regime to manage groundwater use: an application of choice experiment to the Merguellil River Basin. In P. Koundouri (ed.) *The use of economic valuation in environmental policy*. London: Routledge.

Martinez, C.P. and Prantilla, E.B. (2007) Economic valuation of the groundwater in Dumoy aquifer. 10th National Convention on Statistics (NCS), October 1–2, Manilla, Philippines.

Martinez-Paz, J.M. and Perni, A. (2011). Environmental cost of groundwater: A contingent valuation approach, *International Journal of Environmental Resources*, 5(3): 603–12.

McCay, B.J. and Acheson J. M. (eds) (1987) *The question of the commons: The culture and ecology of communal resources*. Tucson, AZ: University of Arizona Press.

McEachern, W.A. (2006) *Economics. A contemporary introduction*. 7th edn. Mason, OH: Thomson South-Western.

McNeill, D. (1998) Water as an economic good, *Natural Resources Forum* 22(4): 253–4.

Menger, C. (1871) *Principles of economics* [*Grundsätze der Volkswirtschaftslehre*], translated by J. Dingwall and B.F. Hoselitz. New York: New York University Press.

Moncur, J.E.T. and Pollock, R P. (1988) Scarcity rents for water: a valuation and pricing model, *Land Economics*, 64(1): 62–72.

Musgrave, R.A. (1956) A multiple theory of budget determination, *Finanzarchiv*, XVII(3): 333–43.

Musgrave, R. A. (1959) *The theory of public finance.* New York: McGraw-Hill.

NAFTA (1992) North American Free Trade Agreement, NAFTA Secretariat, http://www.nafta-sec-alena.org/en/view.aspx?x=343&mtpiID=ALL

NRC (1997) *Valuing ground water: Economic concepts and approaches.* Washington, DC: National Academic Press.

NRC (2002) *The drama of the commons: NRC Proceedings of the Conference on Common Property Resources Management.* Washington, DC: National Academic Press.

Ostrom, E. (1990) *Governing the commons: The evolution of institutions for collective action.* Cambridge: Cambridge University Press.

Ostrom, E. (2008) The challenge of common pool resources. *Environment: Science and Policy for Sustainable Development.* 50(4): 8–21.

Oxford English Dictionary (2007) 2nd edn. Oxford: Clarendon Press.

Perry, C.J., Seckler, D. and Rock, M. (1997) *Water as an economic good: A solution or a problem?* Research Report 14. Winrock: International Irrigation Management Institute; Colombo: International Institute for Agricultural Development.

Plottu, E. and Plottu, P. (2007) The concept of total economic value of environment: A reconsideration within a hierarchical rationality, *Ecological Economics*, 61: 52–61.

Press, J. and Söderqvist, T. (1998) On estimating the benefits of groundwater protection: a contingent valuation study in Milan. In T.M. Swanson and M. Vighi (eds) *Regulating chemical accumulation in the environment.* Cambridge: Cambridge University Press.

Randall, A. and Stoll, J.R. (1983) Existence values in a total valuation framework. In R.D. Rowe and L.G. Chestnut (eds) *Managing air quality and scenic resources at national parks and wilderness areas.* Boulder, CO: Westview Press.

Robbins, L.C. (1932) *An essay on the nature and significance of economic science.* 1st edn. London: Macmillan.

Savenije, H. and van der Zaag, P. (2002) Water as an economic good and demand management paradigms with pitfalls, *Water International*, 27(1): 98–104.

Smith, A. (1776) *An inquiry into the nature and causes into the wealth of nations.* 1st edn. London: Strahan and Cadell.

Stenger-Letheux, A. and Willinger, M. (1998) Preservation value for groundwater quality in a large aquifer: a contingent valuation study of the Alsatian aquifer, *Journal of Environmental Management*, 53: 177–93.

Van Eecke, W. (2006) *An anthology regarding merit goods: The unfinished ethical revolution in economic theory.* West Lafayette, IN: Purdue University Press.

Weisbrod, B. (1964) Collective-consumption services of individual consumption goods, *Quarterly Journal of Economics*, 78: 471–7.

Weiss, E.S. (2005) Water transfers and international law. In E.S. Weiss, L.B. de Chazounes and N. Bernasconi-Osterwalder (eds) *Fresh water and international economic law.* Oxford: Oxford University Press.

Winpenny, J. (1994) *Managing water as an economic resource.* London: Routledge.

5 Water transfers and transferable groundwater rights

This chapter is concerned with water transfers and transferable groundwater rights. It begins with a brief review of water transfers around the world and a description of the types of transfers that have evolved, mainly, in the western USA. It continues with defining transferable groundwater rights, highlighting the subtle distinction from tradable water rights. A discussion follows on groundwater rights as property rights and how groundwater rights may be transferred under different water rights' allocation doctrines. The chapter closes with a discussion on the reasons for making groundwater rights transferable with examples from different countries.

Water transfers

A water transfer usually means the movement of water from one place to another or the reallocation of water among different users, or both. In the first category belong transfers within a basin (intrabasin), outside a basin (transbasin) or across country boundaries (transboundary). These are generally large bulk movements of water involving major engineering works and capital costs. The second category includes the movement of water between sectors, the most common being from agriculture to municipal or industrial supplies, mainly as a means of achieving economic efficiency. There are no large bulk transfers of either surface water or groundwater in this category. They are often temporary, and although they may be large in aggregate, they are rather diffused. Transfers may be enabled by contractual agreements between parties, treaties between governments or legislation allowing for water rights to be transferable.

Generally, water transfers are from areas of ample water resources – which may include fossil groundwater in large groundwater basins – to areas of scarce water resources. Although economic considerations, and recently the participation of the private sector, play an important role, many of the large bulk transfers are essentially non-market transfers in the sense that there is no trading of water or water rights by buyers and sellers. It is true that recipients of the transferred water may have to make an agreed (annual) payment to the donor during the period of the transfer, as for example is the case of Lesotho receiving payment from South Africa, but this is not substantially different from an ordinary consumer paying his or her water bill.

Ancient water transfers

Water transfers are not new. Many cities in antiquity relied for their potable needs on wholesome water conveyed from springs tapped some distance away. The Minoans, more than 4,000 years ago, used aqueducts to bring water to the town of Knossos and the royal palaces from limestone springs; and an aqueduct built by Peisistratus (510 BC) brought water to Athens from the foothill of the Hymetos, 7.5 km from the Acropolis (Mays, 2010). In Mesopotamia, the Assyrian King Sennacherib (705–681 BC) supplied water to Nineveh and the surrounding fields and gardens by canals and aqueducts from a dam on the Tebitu river upstream of the town and from springs and pools at Mount Musri, 50 km away (Tamburrino, 2010). The Romans over a period of more than 500 years (312 BC–226 AD) built 11 aqueducts having a total length of approximately 500 km to supply the increasing water needs of Rome (Hodge, 1991). Some aqueducts conveyed water from springs but many did so from lakes and rivers. At its height, the system supplied $1 \text{ m}^3 \text{ s}^{-1}$ to one million inhabitants.

Examples of modern bulk water transfers in surface water

In recent times water transfer schemes to meet municipal, industrial and agricultural demands have been on a much grander scale. Technology has been a great catalyst. It has enabled large quantities of water to be transferred over long distances through difficult and mountainous terrain, often requiring the construction of large dams, tunnels and pumping stations. Groundwater, as discussed below, has been transferred over distances of several hundred kilometres, by pipelines laid through inhospitable deserts. Generally, transfers provided benefits to the recipients, but at times at the expense of the environment and/or livelihood of the donors. A well known example is the Aral Sea. The diversion of river waters in the period 1960–1987 during the Soviet era to support irrigated agriculture for cotton resulted in a reduction of water inflows with major environmental consequences that have affected both humans and the ecology (McKinney, 2004). The Colorado-Big Thompson project in the western USA is almost a prototype of the interbasin type of water transfer. It is a major scheme, although very much smaller than the Chinese South–North scheme or the proposed Indian National River Linking Project (NRLP). It was first operated in 1947 and transfers surface water over a distance of approximately 400 km across the continental divide from the high mountains of western Colorado to the high plains of eastern Colorado for municipal, irrigation and industrial use (Autobee, 1996). It comprises 13 dams and 10 reservoirs and has a design output of 310,000 acre-feet a^{-1} (0.382 $\text{km}^3 a^{-1}$). The Snowy River scheme in Australia is another large scheme. It started in 1949 and was completed in 1974, and diverts water for irrigation from the Snowy River to the Murray–Darling River basin. The scheme comprises 16 dams, tunnels and aqueducts and has an output of 1.1 $\text{km}^3 a^{-1}$. An undesirable effect has been a reduction in the flows of the Snowy River, which raised concerns over the river ecology and the environment (Ghassemi and White, 2007; WWF, 2007).

A more recent scheme, now nearing completion, is the Olmos irrigation project in northern Peru. It aims to transfer approximately 0.4 km^3 a^{-1} of water via a 19.3 km long tunnel across the Andes from the Huancabamba River in the east to the Olmos plains in the west (H2Olmos, 2010).

China is in the process of constructing one of the largest water transfer projects in the world. The South–North Water Transfer Scheme will take water from the Yangtze basin in the wet south to provide irrigation, municipal (Beijing, Tianjin, and some cities in Hebei, Henan, and Hubei provinces) and industrial supplies to the dry north, where groundwater has been steadily depleted. The project is planned for completion by 2050, when its total diversion capacity will be 45 km^3 a^{-1}. Concerns have been expressed with regard to future water resources availability, and the assessments of water demand especially for irrigation and ecosystems (Liu and Zheng, 2002; Yang and Zehnder, 2005).

The proposed National River Link Project (NRLP) in India, if implemented, will be the largest water transfer scheme in the world. It will divert about 178 km^3 a^{-1} from the potentially water-surplus Himalayan rivers to the water-scarce river basins of western and peninsular India. A strategic analysis of the hydrological, social and ecological issues associated with the project was presented in a workshop in Colombo, Sri Lanka, in 2008 (Amarasinghe and Sharma, 2008).

The Lesotho Highlands Water Project involves transboundary water transfers. It diverts water from the Senqu (Orange) river system in the high rainfall region of the Drakensberg mountains in Lesotho to the Vaal river to supply the dry area of the industrial Gauteng Province of South Africa. The first phase comprises two dams (Katse and Mohale) and a diversion weir (Mutsoku) with a potential yield of 30 m^3 s^{-1} (0.946 km^3 a^{-1}) (LHDA, 2012). Construction started in 1989 and water to South Africa was first delivered in 1998. The project was completed in 2004. It has achieved its objectives of providing water to South Africa and hydroelectric power and revenue to Lesotho, but there has been criticism with regard to corruption, lack of compensation for the impacts on the livelihoods of displaced communities and excessive dependence of the Lesotho economy on the project (Hildyard, 2000; WWF, 2007; Mashinini, 2010).

Examples of modern bulk transfers of groundwater

There are only a few known large bulk interbasin groundwater transfers and no transboundary transfers. The Great-Man-Made-River (GMMR) project in Libya is the largest groundwater transfer scheme in the world. Construction started in 1984 and the two main phases were completed in 2007. Fossil groundwater is extracted from the Nubian Sandstone aquifer in the Sahara desert using five large wellfields comprising more than 1,300 boreholes, each 500–800 m in depth. The scheme transfers via pipelines approximately 2.2 km^3 a^{-1} to the coastal north over distances of 300–600 km. Approximately 70 per cent of the water is used for agriculture, 28 per cent for the municipal supply of Tripoli, Benghazi and Sirte, and 2 per cent for industry. The GMMR is a very ambitious project which utilises beneficially a hidden resource, although still meeting only part of the requirements of the country. The extent of potential effects on the local desert populations and the groundwater supported *sebkhas*, depressions and oases are not yet known. There is also the question of the bequest value of this exhaustible reserve. Under construction is a similar, although much smaller, scheme to transfer 0.1 km^3 a^{-1} from the Southern Desert of Jordan to the capital Amman, 325 km to the north. The scheme, known as the Disi-Mudawarra to Amman Water Conveyance System, is a build-operate-transfer (BOT) project. It commenced in 2009 and is expected to be completed in 2013. It exploits the fossil transboundary Disi-Saq sandstone aquifer and comprises 55 boreholes 500–600 m in depth. There are transboundary issues with Saudi Arabia, the other aquifer state. The long-term impact of the abstraction on the springs issuing into the Dead Sea area is difficult to predict, and, of course, as with the Nubian sandstone aquifer, there is the aspect of bequest value of the resource. Finally, it is worth mentioning the transfer of groundwater to Amman from a wellfield exploiting the basalt and Rijam limestone aquifer system situated north of the Azraq Oasis, 120 km northeast of Amman. Groundwater exploitation began in 1982 at about 0.015–0.022 km^3 a^{-1}. Initially there were no perceived adverse effects but over time it became evident that flows from two springs (Druz and Shishan) were affected: the Druz spring dried up in 1986 and the flow of the Shishan spring decreased by 20 per cent (Naqa, 2010). The Azraq wetland is a RAMSAR conservation site and a reduction in water supply and increases in water salinity threaten its ecological integrity.

The large water transfer projects described above essentially concern the movement of water, usually by the state, to satisfy perceived needs for a particular use at distant locations within or outside a country. The latter are the more difficult to implement, partly because international law in water resources is not particularly robust – for groundwater it does not yet exist – and partly because both the recipient and donor countries have natural fears, the former of dependency and the latter of a possible erosion of sovereignty over its resources. Thus, transboundary transfers take the form of contractual arrangements, as for example between Greece and Cyprus in 2008, which related to transfer by tanker of about 8 Mm^3 to Cyprus to alleviate water shortages at that time (AFP, 2008; Maxim's News Water, 2008); or treaties, such as between Lesotho and South Africa (Treaty on the Lesotho Highlands Water Project, 24 October 2006). Turkey and Israel have been in discussion for several years to provide Israel with 0.05 km^3 a^{-1} over a period of 20 years from the Manavgat River in Turkey. In 2006, the project was halted reportedly due to a rise in costs (Priscoli and Wolf, 2009). However, discussions continued but halted again in June 2010 (*Haaretz*, 2010) following the Gaza Freedom flotilla incident of 31 May, which demonstrates the susceptibility of transboundary transfers to political factors.

Water transfer mechanisms in the western USA

In the arid western USA, water transfers have been taking place for more than 100 years. Several types have evolved which have been described by various authors

(NRC, 1992; Eheart and Lund, 1996; California Environmental Protection Agency (CEPA), 1999; WTW, 2002; Tarlock, 2005). A summary is presented below:

1   *Permanent transfers* These involve the acquisition of water rights and a change in ownership of the right.
2   *Contingent transfers or dry year options* These are temporary transfers contingent on meeting short-term water shortages, such as during drought. They may take the form of long-term or contingent leasing contracts covering several decades, intermediate-term contracts (3–10 years) or short-term (1–2 years).
3   *Spot market transfers or leases* These are short-term transfers or leases agreed on between a water-right owner and a new user to use a fixed quantity of water over a specific period of time. Farmers or cities use this mechanism to temporarily supplement their water supplies during dry periods.
4   *Water banks* This is a formal mechanism sanctioned by the state for pooling surplus water rights for rental to others. Water banks are operated by a central banker. Users sell water to the bank for a fixed price and buy water from the bank at a higher price; the difference in prices typically covers the bank's administrative and technical costs. The functions of water banks vary but their main purpose is to facilitate the legal transfer and market exchange of surface water, groundwater and storage entitlements; other services may include acting as intermediaries or brokers, bringing together buyers and sellers, or acting as a clearinghouse or market maker (Clifford *et al.*, 2004) .
5   *Water wheeling or water exchanges.* Water wheeling is similar to electric power being wheeled or exchanged through the transmission system between power companies and electric power plants, to make it less expensive and more reliable. For water, wheeling or exchange can be applied through the water-conveyance system and storage facilities.
6   *Transfer of reclaimed (salvaged), conserved and surplus water* This is not strictly speaking a form of water transfer. It involves providing incentives to users, including rebates to customers to use less water and, reducing water losses during distribution or conveyance through measures, such as financing irrigation improvements to save water in exchange for rights to use the water.
7   *Storage of surplus surface water* during wet years in aquifers (artificial recharge) allowing the use of more groundwater during dry years, or providing surplus surface water to irrigators in exchange for resting groundwater and allowing storage to recover for use during drought years.

Informal water markets

In informal water markets, trading involves the transfer of water volumes between willing buyers and sellers. Water is sold by farmers to other farmers, to public water supply companies, such as in Chennai, India, or to private consumers in urban areas. Unlike formal water markets, informal markets do not involve the exchange of legally defined water rights. Informal water markets have been important in South Asia (Shah, 1993). In Bangladesh, 88 per cent of well owners sell a proportion of their groundwater to other farmers, in Nepal Terai 62 per cent; in east India 46 per cent; and in Pakistan Punjab 33 per cent (Mukherji and Shah, 2005). As groundwater transfers are unregulated, informal markets have led to over-abstraction, as farmers abstract groundwater to satisfy both their own needs and for selling to others. In the state of Gujarat, India, Shah *et al.* (2008) reported negative effects on the poor due to groundwater scarcity. However, there has also been evidence of improved water-use efficiency and increased equity in resource access (Manjunatha *et al.*, 2011). Poor farmers have generally benefited as they are able to achieve higher agricultural productivity by buying irrigation water. In India, large farmers generally sell groundwater to poor farmers (Purushottam and Sharma, 2006) who do not have the money to drill and equip their own wells. This is not the case in other countries, for example Australia, where poor farmers sell their water (rights) to larger farmers (Bjornlund, 2006), or Chile where indigenous communities have sold their water rights to privatised water companies, mining corporations and agri-businesses (see Chapter 6, section on Chile).

Transferable and tradable water rights

In general, writers on water rights have used the terms 'transferable' and 'tradable' interchangeably, suggesting that they take them to mean more or less the same thing. The Environment Agency (EA) of England and Wales defined a 'water rights trade' as 'the transfer of licensable water rights from one party to another, for benefit' (EA, 2003). Tarlock (2005) notes that the term 'water transfer' is also often used interchangeably with water marketing. But he points out that marketing 'technically refers only to the sale or donation of an existing entitlement', whilst water law 'has reserved the term "transfer" for changes in use or point of use of an existing entitlement'. Hodgson (2006) combined the two terms as 'tradable transferable' water rights. In this book, the term 'transferable water right' or 'transferable groundwater right' is understood to mean a water right that may be transferred independently of the parcel of land on which water is used. The removal of the legal requirement for a water right to be transferred independently of land provides the opportunity for trading for benefit, which is usually money, of water rights themselves. Thus, 'tradable water rights' are the potential consequence of water rights becoming transferable. Following from this, a circumstance may be envisaged in which water rights may be transferred from one party to another not as part of a trade, but for example, for the benefit of the resource.

Water rights as property rights

Whether or not water rights are property rights is important when considering their transferability without involving the parcel of land over or under which the water occurs. According to Sax (1990) water rights are property, and not a special kind of property right which cannot be regulated in the same manner as other property rights. He considers that from a constitutional (US) perspective all property rights have exactly the same status, although the definition of a particular property right may make it something less than a full fee simple interest (a full fee simple interest is one that is permanent and absolute with freedom to dispose of it at will). For example the right may be defeasible (a fee simple defeasible interest is one that has conditions attached to it) or of a limited term. He points out that water rights have no higher or more protected status than any other sort of property right and in fact they have less protection than most other property rights. Dellapenna (2000) considers that 'the paradigm of property in common law remains the fee simple absolute' and that 'flowing water, like any ambient source, simply does not fit easily into such a paradigm'. He contends that in the USA no marketable scheme of water rights the same as would apply to land has ever actually been implemented, and that the major changes in private property rights in water have stressed the limitations that the public nature of water resources impose upon private rights. According to Hodgson (2004) the fact that water rights gain their existence from an administrative or regulatory procedure does not by itself preclude them from being property rights. He notes, however, that water rights created under public or administrative law do not have all of the attributes of private property rights. Tarlock (2005) considers that water rights are a form of property rights, although different from land rights and other forms of property, and in this respect the general presumption of free alienability applies.

From the above, the consensus appears to be that although water cannot be possessed in the same way as land, and differs from it in other ways, water rights are a form of property right, although not having the attributes of private property rights. Hodgson (2004) provides a detailed comparison between land-tenure rights and land rights dealing with aspects of security of right, the substance of the right, charging, international law, and tradability. Some of the differences (not all from Hodgson) between water rights and land rights are the following:

1   Land is fixed whereas water, either as surface water or groundwater, is flowing. Thus, water as it occurs in nature cannot be possessed or disposed off in the same way as land.
2   There is constancy in land in terms of its size and position, and therefore, land rights are secure and easy to measure. Groundwater and surface water on the other hand, being largely dependent on rainfall, are temporally variable and inherently less secure. Moreover, groundwater given its occurrence underground is difficult to measure.
3   The use of groundwater, particularly its intensive use in recent times, leads to externalities with regard to impacts on the environment, resource availability and other users. There are also externalities in land use but possibly not to the same extent or degree. Water rights systems have sought to minimise such impacts through the imposition of conditions and protection measures.
4   The duration of land ownership is normally for an indefinite period but even time limited tenures are acceptable and respected. Traditional water rights of unlimited duration are quickly disappearing and are steadily being replaced by time-limited rights.

5   Unlike land, groundwater is not an exclusive resource, partly because of its importance to human life and the environment, and partly because it would be difficult to exclude others from using it (see Chapter 4 for a more detailed discussion on this).

Water rights transfers under different allocation doctrines

The various doctrines that govern the allocation of groundwater rights have been described at some length in Chapter 3. In the great majority of countries, water rights are still attached to land and can only be transferred with the transfer of the parcel of land to which they relate.

Riparian rights are ordinarily transferred when a landowner sells his or her land. Tarlock (2005) points out that transfers of riparian rights remain problematical, as a riparian does not have a right to a fixed amount of water, until the right has been adjudicated. Even so, the right may theoretically still be vulnerable to the claims of others who were not parties to the adjudication.

Under the prior appropriation doctrine, water rights are transferable so long as no harm is suffered by other appropriators. However, the doctrine in its aim to protect (senior) appropriators who use water mainly for agriculture, does not encourage the transfer of water to other uses (Huffaker *et al.*, 2000). Also, transfers are subject to various limitations which vary from state to state. Thus, although transbasin and interbasin transfers are generally permitted, in some states legislation has been enacted modifying the prior appropriation doctrine so as to protect the equities and interests of the area of origin and prevent adverse effects on economic and ecological aspects (Getches, 2008). Additional criteria to evaluate transfers involving public interest and social welfare have also been put forward by stakeholders who do not own water rights, which weaken further the appropriation doctrine in relation to water rights transfers (Gardner, 2003). There are also limitations to the transfer of water rights when changes of use may harm other appropriators, including junior appropriators, as a result of the change. Finally, there is a limitation on the quantity of the water transfer, which is based on the (lesser) quantity that has been actually used in the past and not on the quantity prescribed in the right. Estimates of historic use may not always be straightforward, and can result in high costs in experts and lawyers fees. Actual use may be determined from records, although in most cases it is likely that historic measurements are lacking. Indirect methods can include area irrigated and crop water requirements, dimensions of diversion structures, capacity of pumps and number of hours of daily pumping, spot flow measurements, enquiries from users and neighbours. This last method must be looked at carefully, as memories are not always reliable and could be biased by individuals' motives.

Under the reasonable use doctrine, water rights cannot be severed from the land, and therefore, the transfer of groundwater rights separate from the land is not allowed. Also, groundwater may not be transferred off the land on which it is pumped if others are injured. In correlative rights, water rights are transferable, but subject to the use limitation for non-overlying land, i.e. permitted only if there is surplus of water available or the surplus is not needed by overlying landowners. Under the absolute ownership doctrine (rule of capture in Texas) in which the landowner owns the groundwater beneath his land, water rights are freely transferable.

Reasons for introducing transferable water rights

Transferable water rights arose from concerns about scarcity influenced by the political dogma of the free market. As mentioned in Chapter 4, the argument for the introduction of transferable water rights rests largely on considerations of economic efficiency and water achieving its economic value. For groundwater, it mainly concerns the transfer of water from a low-value high-volume use in irrigated agriculture to a high-value low-volume urban public supply or industrial uses. This is the case in many developing countries, which have a strong and often subsidised irrigated agricultural sector, and an increasing public water demand, as people move into urban areas.

Assigning property rights to water rights enables their transfer and trading independently of land ownership. The freeing of water rights from land rights enables a more flexible allocation of water resources while water rights trading potentially reduces wastage and helps in using water resources more sustainably.

Transferable water rights in different countries

To date transferable water rights have been applied only in a handful of countries, the earliest application in the western USA following from the prior appropriation doctrine. In the last 30 years transferable water rights have been introduced in Chile (1981), Mexico (1992), Australia (1998) and England and Wales (2003), and the experiences in these countries are discussed in Chapter 6. The reasons why these and other countries have allowed the transfer and trading of water rights in their legislative codes vary a little, but all share the same objective of treating water as an economic good. Underlying this have been the politics of the free market which held sway in the 1980s and 1990s and still is in evidence today. A good example of this is Chile where changes in water legislation followed the removal of the socialist government of Allende in 1973 and its replacement by the right-wing military junta of General Pinochet. The 1981 Water Code in Chile aimed at strengthening the free market economic model and private-property water rights, reducing the role of government, encouraging inter-sectoral transfers and trading of water rights and achieving the economic value of water. Similarly, in Mexico, there has been a general trend towards water privatisation, which was further stimulated by Mexico's membership of NAFTA from 2000. The 1992 National Water Law aimed at decentralizing the authority to grant water concessions – which until then could only be issued by the president – properly addressing management issues, and encouraging water transfer and trading to achieve economic efficiency. In Australia, water codes in different states in the last 10–20 years have encouraged the conversion of existing licences (or entitlements) to transferable water allocations with the objective of moving away from land-based water rights legislation in order to better manage water resources, prevent and control overexploitation, ensure environmentally sustainable development, and recognise the economic value of water. The Productivity Commission of the Commonwealth of Australia (PCCA, 2003) considered that if water rights are tied to land title or use conditions the flexibility and efficiency of reallocation of the resource through trading is diminished. In England and Wales privatisation of water was part of a political initiative that started in the late 1980s, and included the privatisation of many of the major industries of the country. The 2003 Water Act made abstraction licences tradable, which simplified the licence transfer process, such that there was no longer a need for a change in the occupation of land for a licence transfer to occur. Also, the complexity of transferring a licence when the land associated with the licence was divided for sale to different individuals was removed. Two of the reasons for encouraging water rights trading were to achieve a more efficient allocation of water rights in England and Wales and to protect the environment. The third was to meet the expectations expressed by government in 'Tuning Water Taking' which related to decisions on the use of economic instruments to water abstraction (EA, 2003). It was considered that placing a tradable value on water resources would encourage water rights to move to the person who placed the highest value on them, which would promote economic efficiency. To date, water rights trading has not been particularly evident (see Chapter 6). Further reforms to encourage market participation are proposed in the very recent (July 2012) Draft Water Bill.

Spain, South Africa, Armenia, and China have also responded to the ideas of the free market as providing the means for reallocating water resources by including in their water codes provisions enabling the transferability of water rights.

Spain in its reform Act 46/1999 of the 1985 water code introduced complex legislation allowing the transfers of water-use rights and envisaging the formation of water exchange centres, similar to the California water banks (Embid, 2002). Trading of water rights is only allowed for publicly owned water. Privately owned water rights cannot be traded unless they become publicly owned. Water rights trades must be approved by the relevant river basin authority (RBA) and interbasin transfers by the Ministry of the Environment, Rural and Marine Affairs. Water exchange centres are state-run bodies authorised by the Council of Ministers and set up by the RBAs.

In South Africa, the 1998 National Water Act allows for the transfer of water use for another purpose or to another land by the same person if this is authorised by a water management institution. National water strategy provides for inter-catchment water transfers between surplus water management areas and deficit water management areas. The government has ultimate responsibility for the management and protection of water resources.

Armenia in its 2002 water code allows water use permit holders to sell or otherwise transfer a portion of their permitted water right to a third party, subject to the procedures laid down by government and providing their permit does not specifically prohibit the transfer.

In China, the water legal system has evolved in the last 20–30 years, with the enactment of the 1988 Water Law and its revision in 2002. A comprehensive description of recent Chinese water law is to be found in Wouters *et al.* (2004). The 2002 Water Act attaches importance to strategic planning and provides the framework for the integrated management of water resources (Speed, 2009; Shugang, 2010). Water rights became transferable in 2006 under article 27 (State Council 26) of the regulation on water withdrawal permit and collection of water resource fee. The article refers to the transfer of water that has been saved as a result of technological improvements (Shugang, 2010), which is similar to the western US mechanism of 'salvage' or 'conserved' water. State Council 26 establishes an initial water rights distribution and water rights transfer system (Wouters *et al.*, 2004).

In none of these countries has there been any significant transfer or trading of groundwater rights. In Spain, informal trading of water is common in agriculture among farmers involving temporary transfers within an irrigation community (OECD, 2010). Formal trading of water rights remains rare. For example, in the Segura basin it amounted to less than one per cent of total consumption in 2001–2005 (Garrido and Calatrava, 2010). Water exchange centres have been set up by some RBAs (Júcar, Segura, Guadiana and Guadalquivir basins (Embid, 2012)) in order to encourage seasonal trades of water rights during periods of drought. RBAs have purchased water rights to meet environmental objectives. According to Garrido *et al.* (2012) water rights trading for the period 2005–2008 has been limited, patchy and discontinuous (mainly during the last drought) with few buyers and sellers. There has been strong state participation in promoting and facilitating water rights exchanges. In the Guadiana basin the Guadiana Exchange Centre bought land and the associated groundwater rights in order to arrest the continued decline of groundwater levels due to irrigation abstractions and the consequent effects on the Tablas de Daimiel wetland. In 2008, the economic crisis has a put a stop to further purchases. Overall, water rights trading has led to economic efficiency (i.e. buyers used water at a higher economic value than sellers) and environmental gains through state purchases of water rights. There have been, however, critics who expressed fears that water rights trading may eventually lead to the privatisation of a publicly owned resource and concerns of financial benefits by traders who benefited from a freely allocated subsidised resource and from exemptions from the payments fees for the use of aqueducts (Garrido *et al.*, 2012).

In South Africa, there have been only a few transfers among farmers of unused surface water (sleeper rights), such as on the Lower Orange River (Nieuwouldt and Armitage, 2004). In China, there have been regional water transfers, such as the Dongyang–Yiwu scheme, transfers of water saved from linings of irrigation canals and transfers between farmers trading their water tickets (an annual volume of water purchased from the district or water users association) (Speed, 2009). Although water tickets are freely tradable, in practice there are only few instances of trading, mainly between neighbours. It seems, however, that there has been considerable government-managed reallocation of surface waters stored in reservoirs from irrigation to domestic, industrial and tourist uses. According to Jiang *et al.* (2012), agricultural water use has decreased from 70 per cent in 1997 to about 62 per cent in 2008. The lack of coordination between departments and lack of clarity in water rights ownership cause problems of allocation, and potential hardships to farmers.

Concluding remarks

Water transfers have been taking place since ancient times. Modern engineering technology has made it possible for large volumes of water to be transferred over long distances across mountains, through deserts and between countries. Recipients have generally enjoyed the benefits of ample water supplies but donors have, at times, suffered from a reduction in stream flows, impacts on livelihoods and adverse effects on the environment. Bulk water transfers have mainly involved surface water. There are very few examples of bulk groundwater transfers, mainly these are from fossil aquifers in the Middle East and North Africa. Invariably all large water transfers have been undertaken or been sponsored by the state.

Legislative changes in the water codes of some countries, partly motivated by political ideology and partly by economic efficiency considerations, have enabled groundwater rights to be treated as property rights and to be transferred independently of land. The added flexibility in water resources allocation should have, theoretically, encouraged groundwater rights trading and through this its reallocation to higher value uses, and exploitation at a more sustainable level. Some experience in transferable groundwater rights has been gained in the western USA, Chile, Mexico, Australia, and England and Wales, and detailed descriptions and conclusions are presented in Chapter 6. Other countries which have introduced water rights transferability in their water codes are: Spain, South Africa, Armenia, and China. Transfers in these countries are still being developed mainly through state supported schemes.

References

AFP (2008) Drought-hit Cyprus signs deal to ferry water from Greece, Agence France-Presse (AFP), 21 April 2008, Nicosia, Cyprus.

Amarasinghe, U.A. and Sharma, B.R. (eds) (2008) Strategic analyses of the National River Linking Project (NRLP) of India, Series 2. *Proceedings of the workshop on analyses of hydrological, social and ecological issues of the NRLP.* Colombo, Sri Lanka: International Water Management Institute.

Autobee, R. (1996) *Colorado–Big Thompson Project,* Denver, CO: US Bureau of Reclamation. http://www.usbr.gov/projects/

Bjornlund, H. (2006) Can water markets assist irrigators managing increased supply risk? Some Australian experiences. *Water International* 31(2): 221–32.

CEPA (1999) *A guide to water transfers.* Sacramento, CA: California Division of Water Rights, State Water Resources Control, State of California. http://www.waterrights.ca.gov

Clifford, P., Landry, C. and Larsen-Hayden, A. (2004) *Analysis of water banks in the western states.* Washington, DC: Department of Ecology.

Dellapenna, J. W. (2000) The importance of getting names right: The myth of markets for water. *William and Mary Environmental Law and Policy Review*, 25: 317–77.

EA (2003) *Trading water right: A consultation document.* Bristol: Environment Agency of England and Wales.

Eheart, J.W. and Lund, J.R. (1996) Water-use management: Permit and water-transfer systems. In L.W. Mays (ed.) *Water resources manual handbook.* New York: McGraw-Hill.

Embid, A. (2002) The evolution of water law and policy in Spain, *International Journal of Water Resources Development*, 18(2): 261–83.

Embid, A. (2012) Legal reforms that facilitate trading of water use rights in Spain. In J. Maestu (ed.) *Water trading and global water scarcity international experiences.* Washington, DC: RFF Press/Routledge.

Gardner, B.D. (2003) Weakening water rights and efficient transfers, *International Journal of Water Resources Development*, 19(3): 7–19.

Garrido, A. and Calatrava, J. (2010) Trends in water pricing and markets. In A. Garrido and M .Llamas (eds) *Water policy in Spain.* Boca Raton, FL: CRC Press.

Garrido, A., Maestu, J., Gomez-Ramos, A., Estrela, T., Calatrava, J., Arrojo P. and Cubillo, F. (2012) Voluntary water trading in Spain: a mixed approach of public and private initiatives. In J. Maestu (ed.) *Water trading and global water scarcity international experiences.* Washington, DC: RFF Press/Routledge.

Getches, D.H. (2008) *Water law in a nutshell: Overview and introduction to water law*, 4th edn. St Paul, MN: Thomson-West.

Ghassemi, F. and White, I. (2007) *Inter-basin water transfers: Case studies from Australia, United States, Canada, China, and India.* International Hydrology Series. Cambridge: Cambridge University Press.

H2Olmos (2010) Olmos Irrigation Project. http://en.h2olmos.com/olmos-irrigation-project.html

Haaretz (2010) Turkey halts all state energy and water projects with Israel, June 4, 2010, English Edition, Tel Aviv, Haifa.

Hildyard, N. (2000) The Lesotho Highland Water Development Project: What went wrong? Paper presented at 'Corruption in Southern Africa: Sources and Solutions' conference, July 10 2000, Chatham House, London. http://www.globalpolicy.org/nations/corrupt/lesotho.htm.

Hodge, A.T. (1991) *Roman aqueducts and water supply.* 2nd edn. London: Duckworth.

Hodgson, S. (2004) *Land and water: The rights interface.* FAO Legislative Study 84 Rome: Food and Agriculture Organisation.

Hodgson, S. (2006) *Modern water rights theory and practice*, FAO Legislative Study 92. Rome: Food and Agriculture Organisation.

Huffaker, R., Whittlesey, N. and Hamilton, J.R. (2000) The role of prior appropriation in allocating water resources into the 21st century, *International Journal of Water Resources Development* , 16(2): 265–73.

Jiang, Y., Luo, Y, Peng, S., Wang, W. and Jiao, X. (2012) Agricultural water transfers in China: Current issues and perspectives, *Procedia Engineering*, 28: 36367.

LHDA (2012) Engineering components. Lesotho Highlands Water Project. http://www.lhda.org.ls/engineering/default.htm

Liu, C. and Zheng, H. (2002) South-to-north water schemes for China, *Water Resources Development*, 18(3): 453–71.

Manjunatha, A.V., Speelman, S., Chandrakanth, M.G. and Van Huylenbroeck, G. (2011) Impact of groundwater markets in India on water use efficiency: A data envelopment analysis approach, *Journal of Environmental Management*, 92(11): 2924–9.

Mashinini, V. (2010) *The Lesotho Highlands Water Project and sustainable livelihoods, policy implications for SADC (South African Development Community)*, Briefing No 22. Johannesburg, Africa: Institute of South Africa.

Maxim's News Water (2008) Greek ships bring drinking water to Cyprus, Maxim's News Network, United Nations, 9 July 2008.

Mays, L.W. (2010) A brief history of water technology during antiquity before the Romans. In L.W. Mays *Ancient water technology*. Dordrecht: Springer.

McKinney, D.C. (2004) Cooperative management of transboundary water resources in Central Asia. In D. Burghart, D. and T. Sabonis-Helf (eds) *The tracks of Tamerlane: Central Asia's path into the 21st century*. Washington, DC: National Defense University Press.

Mukherji, A. and Shah, T. (2005) Socio-ecology of groundwater irrigation in South Asia: An overview of issues and evidence. In A. Sahuquillo (ed.) *Groundwater Intensive Use*. Leiden: Taylor & Francis.

Nieuwouldt, W.L. and Armitage, R.M. (2004) Water market transfers in South Africa: Two case studies, *Water Resources Research*, 40: W09S05.

NRC (1992) *Water transfers in the West: Efficiency, equity, and the environment*. Washington, DC: National Research Council, National Academy Press.

Naqa, El. A. (2010) *Study of salt water intrusion in the Upper Aquifer in Azraq, Jordan*. Gland: International Union for Conservation of Nature (IUCN).

OECD (2010) *Spain*, Economic Surveys 2010/9. Paris: OECD publishing.

Priscoli, J.D. and Wolf, A.T. (2009) *Managing and transforming water conflicts*. Cambridge: Cambridge University Press.

PCCA (2003) *Water rights arrangements in Australia and overseas*. Commission Research Paper. Melbourne: Productivity Commission of Australia.

Purushottam, S. and Sharma, R.C. (2006) Factors determining farmers' decision for buying irrigation water: Study of groundwater markets in Rajasthan. *Agricultural Economics Research Review* 19(1): 39–56.

Sax, J.L. (1990) The constitution, property rights and the future of water law, *University of Colorado Law Review*, 257: 257–82.

Shah, T. (1993) *Groundwater markets and irrigation development: Political economy and practical policy*. Bombay: Oxford University Press.

Shah, T., Bhatt, S., Shah, R.K., Talati, J. (2008) Groundwater governance through electricity supply management: assessing an innovative intervention in Gujarat, Western India. *Agricultural Water Management* 95: 1233–42.

Shugang, P. (2010) China's legal system for water management: Basic challenges and policy recommendations, *International Journal of Water Resources Development*, 26(1): 3–22.

Speed, R. (2009) Transferring and trading water rights in the Peoples' Republic of China, *International Journal of Water Resources Development*, 25(2) 269–81.

Tamburrino, A. (2010) Water technology in ancient Mesopotamia. In L.W. Mays (ed.) *Ancient Water Technologies*. Dordrecht: Springer.

Tarlock, A.D. (2005) Water transfers: A means to achieve sustainable water use. In E.D. Weiss, L.B. de Chazournes and N. Bernasconi-Osterwalder (eds) *Fresh water and international economic law*, International Economic Law Series. Oxford: Oxford University Press.

Wouters, P., Hu, D., Zhang, J., Tarlock, D. and Andrews-Speed, P. (2004) The new development of water law in China, *University of Denver Water Law Review*, 7(2): 243–308.

WTW (2002) *Water transfer issues in California, Final report to the California State Water Resources Control Board by the Water Transfer Workgroup (WTW)*, June 2002. Sacramento, CA: WTW.

WWF (2007) *Pipedreams? Interbasin water transfers and water shortages*, Zeist: WWF Global Water Programme. http://www.wwf.or.jp/activities/lib/pdf/pipedreams_27_june_2007_1.pdf

Yang, H. and Zehnder, A.J.B. (2005) The south–north water transfer project in China: An analysis of water demand uncertainty and environmental objective in decision making, *Water International*, 30(3) 339–49.

6 International experience in the implementation of transferable groundwater rights

This chapter provides a detailed review based mainly on published information relating to the experience with transferable groundwater rights in five countries: Chile, Mexico, western USA, Australia, and England and Wales. Simplified administrative maps of each country are presented at the start of each relevant section. The review covers the following main areas: overview of water resources and their exploitation, the legislative framework, the extent of groundwater transfers and trading, impediments to transferable groundwater rights, benefits and negative externalities.

It is recognised that the coverage of five countries does not quite constitute an 'international experience'. Nevertheless, each of the countries examined has a sufficiently different geographic, climatic, legal, socio-political and economic background to provide information which should be useful to other similar countries. In Asia, where much of the globe's irrigation occurs, informal water markets have developed (these are briefly described in Chapter 5 and are not the subject of this book) but examples of formal water rights trading are lacking. However, parallels may be drawn on the impacts on poor farming communities in Asia from the experience of similar communities in Chile and Mexico.

Finally, it must be pointed out that, in general, the experience in groundwater rights transfers, especially in large volumes across sectors, is fairly limited, and not as widespread as for surface water rights. However, surface water rights transfers provide valuable lessons and insights which may be helpful to the consideration of groundwater.

Chile

Water resources

Chile stretches about 4,000 km north to south, along the western part of South America. It is 375 km at its widest and 90 km at its narrowest and bounded by the Andes Mountains to the east and the Pacific Ocean to the west. Forty per cent of the population of approximately 15.3 million live in the Santiago metropolitan area (2002 population census). Before 2006, Chile was divided into 12 administrative regions, numbered from north to south in Roman numerals I

Figure 6.1 Administrative map of Chile

to XII, except for the Santiago Metropolitan Region, which was designated RM (Región Metropolitana). In 2007, two regions were added: XV in the north by subdividing Region I; and XIV in the south by subdividing Region X. In the discussion that follows the old subdivision is used. The northern part of Chile is arid to semi-arid and includes the famous Atacama Desert, where rain is meagre, if it falls at all. However, the area is rich in mineral deposits, especially copper, and it is where Chile's mining activity largely occurs. As a result, it is in this region that demand for water has become important in recent years. A comprehensive summary of the water resources of Chile has been provided by Salazar (2003). In general, precipitation and runoff increase southwards. Thus, the average annual rainfall in the north (Regions I and II) is 59 mm, in the middle (Regions III to X) 1,246 mm, and in the south (Regions XI and XII) 2,363 mm. Similarly, the mean surface water runoff is 21 $m^3 s^{-1}$ in the north, 9,130 $m^3 s^{-1}$ in the middle and 20,260 $m^3 s^{-1}$ in the south. Annual evaporation is lowest in the arid north, 55 mm, increasing to 377 mm in the middle and 408 mm in the south. The flows of rivers also increase southwards. Thus, the median flow of the Rio Loa in Region I in the north is 2.8 $m^3 s^{-1}$, the Rio Limari in Region IV in the middle 15. 1 $m^3 s^{-1}$, the Rio Maipo in Metropolitan Santiago 116.0 $m^3 s^{-1}$ and the Rio Bio Bio in Region VII in the south 639 $m^3 s^{-1}$. Minimum flows of the northern rivers can be very low ranging from zero to less than 100 $l s^{-1}$. Surface water use amounts to 674 $m^3 s^{-1}$, of which approximately 81 per cent is used for irrigation, just over 4 per cent for potable supply, 8 per cent goes to industry, and 7 per cent to the mining sector. Non-consumptive hydroelectric use amounts to 1,603 m^3 s^{-1}.

The groundwater resources of Chile are not well known, partly because of their occurrence in discrete valley systems, often of tectonic origin. Although groundwater is not as abundant as surface water, it is nevertheless important as a source of water for irrigation and domestic supply for small farmers, and more recently for mining operations especially in the north of the country. The capital Santiago also relies on groundwater for part of its needs. Groundwater use amounts to approximately 88 m^3 s^{-1}, which is about 12 per cent of surface water use. Of this 49 per cent is used for irrigation, 35 per cent for potable supply, and 16 per cent for industry (Salazar, 2003).

In the north, groundwater occurs in thick Quaternary alluvial deposits. It is generally of fossil origin, although it is probable that recharge events have been occurring over the last 100 to 1,000 years (Houston and Hart, 2004). In general, groundwater recharge in the northern areas is very small or negligible, increasing in the southern areas where rainfall is more plentiful. Average recharge for the whole country has been given by Salazar (2003) as approximately 215 m^3 s^{-1}, which is significantly greater than the average usage of about 88 m^3 s^{-1}. This may be a little misleading as much of the groundwater exploitation occurs in the areas of little or no recharge. Thus, in the Santiago Metropolitan Region, groundwater recharge is about 55 m^3 s^{-1} and the actual abstraction about 60 m^3 s^{-1}. Exploitation of the thick upper Mapocho alluvial aquifer by deep boreholes in the 1990s to supply a newly developed affluent area of Santiago led to a continuous decline of groundwater levels, which strongly indicated that abstraction was significantly greater than recharge (author's experience). In the last few years, overexploitation of groundwater resources has led the DGA (Dirección General de Aguas) to designate an increasing number of protected areas where further exploitation is not permitted.

Water legislation

Regarding water legislation, the 1951 Water Code systematised existing water rights and introduced public regulation which was strengthened by the 1967 Water Code. In the period 1973 to 1980 constitutional changes to the right led to the 1981 Water Code (Decreto Con Fuerza de Ley 1.122. Fija Texto del Codigo de Aguas, Publicado Oficial de Chile 29.10.81) which limited government control on water resources, assigned private property rights to water rights and encouraged free markets. Water rights were separated from land title, made transferable and divisible and issued in perpetuity. Water disputes were to be resolved by ordinary courts. There has been no consultation with stakeholders either before or during the implementation of the Code.

Water rights transfers

Much has been written about water rights trading in Chile, and a good description has been presented by Bauer (2004). In general actual data on numbers of transactions and volumes traded have been lacking, most authors preferring, especially in the 1990s, to debate the theoretical and ideological aspects of water rights markets, and conclude favourably or not, depending on their beliefs, political bias or the organisations they represented. The registration of water rights has been generally slow. However, based on the DGA (Dirección de General de Aguas) website, the registered number of groundwater rights appears to have accelerated in the last few years. In the three-year period between 2006 and 2009, approximately 19,000 rights were registered in comparison to less than 4,000 in the previous three years. Salazar (2003) indicated that the volume of authorised groundwater rights was about 107 m^3 s^{-1}, or about twice the recharge to the aquifers. In the upper Santiago Valley aquifer, an evaluation of the resource by Muñoz (2003) indicated that only 67 per cent of the water rights allocation can be sustainably extracted.

It is difficult to establish the proportion or volumes of traded groundwater rights in comparison with the total available groundwater resources countrywide or even within individual basins. Nevertheless, it appears that the proportion of groundwater trades nationwide has been rather small.

According to Global Water Intelligence (GWI, 2010) trading activity in Chilean water rights has been sporadic, but growing. In 2008, market activity was concentrated in a few areas in the centre and Metropolitan Santiago. There was little or no marketing in the south. There are no homogeneous prices for water rights, which can vary even within the same sector and river basin. Prices for groundwater rights are higher than for surface water, reflecting the greater certainty of groundwater during drought. In the Atacama region the price of groundwater rights is at least US$ 50,000 per l s^{-1}. In the upper section of the Mapocho Alto aquifer in the upper part of Santiago, prices reach US$ 25,000 per l s^{-1}, and in the central part of Metropolitan Santiago where there is greater availability and more buyers and sellers US$ 8,000–10,000 per l s^{-1} (GWI, 2010).

The Limari River basin in north-central Chile (Region IV) appears to be the prime example of an area in which water rights trading has been active, comprising both temporary and permanent trades. Local real estate agents broker and facilitate transactions. The water market involves surface water largely for irrigation within the agricultural sector. Bauer (2004) attributes the success of the Limari water market to factors other than the Water Code, namely: (a) the existence of an adequate water storage capacity from three reservoirs built for irrigation in the 1930s and 1970s by government and maintained by the Ministry of Public Works; (b) well-organised local water users associations ensuring that irrigation canals are well maintained; and (c) a climate that favours the growing of high value fruits for export. In the period 1994 to 1996, the price per cubic metre of water in the Paloma system of Rio Limari, increased from around $0.01 to $0.16, while the volume of storage decreased from 1,000 Mm^3 to less than 300 Mm^3 (Salazar, 2003). The Maipo sub-catchment in the higher Mapocho River, which includes Metropolitan Santiago, is an area where water rights trading has involved transactions not only within the irrigated agricultural sector but also by urban water supply companies or estate developers buying water rights from farmers.

In this area, 40 per cent of the agricultural rights have been sold to public water supply (Salazar, 2003). However, according to Bauer (2004), only a small number of water rights are sold each year, and although some reallocation from irrigation to public supply has taken place prices vary widely. In the north of the country, mining companies have purchased water rights from agriculture and indigenous populations at prices generally varying from US$ 75,000 to US$ 225,000 per l s^{-1}. In one case, a volume of 630 l s^{-1} was purchased at US$ 136 million (Arrueste, 2008). These seemingly high purchase prices have to be seen in the context of the intended life of the mining operation and, of course, where replenishment is lacking, the ability of the groundwater resource to sustain the required quantity over the required period. At the higher purchase cost of US$ 225,000 per l s^{-1}, the corresponding average cost per cubic metre after ten years of production will be approximately US$ 0.7, which is not very different from public water supply prices to consumers. Nevertheless, growing social pressures, water resources scarcity and a reduction in authorised water rights are forcing mining companies to consider other sources of water, such as desalination (Arrueste, 2008). As the demand for groundwater in the mining sector of the arid areas in the north has increased, the DGA has been refusing to grant new water rights on the basis that hydrological data do not indicate the availability of sufficient groundwater.

In the mid-1990s most researchers and authors (Gazmuri and Rosegrant,1994; Rosegrant and Binswanger, 1994; Rosegrant and Gazmuri, 1994; Peña, 1996), especially those associated with the World Bank, considered that the trading of water rights in Chile was a success, and one to be emulated by other Latin American countries (World Bank, 1994). This assertion was challenged by Bauer in a number of papers arising from his field research in Chile for a PhD degree, culminating in his book *Siren song* (Bauer, 2004). Using quantitative and qualitative evidence, he came to the view that water rights transactions were quite uncommon in most parts of Chile, and as a general rule, Chilean water markets were relatively inactive. He also found that the great majority of water rights transactions took place within the agricultural sector and did not involve non-agricultural uses. He attributed this lack of success of water markets to a number of limiting factors, including: the physical constraints of geography in Chile which make inter-basin transfers expensive, a rigid or inadequate water supply infrastructure, and uncertainty and confusion about water titles. He also noted that there was a cultural and psychological resistance to the treatment of water as a commodity, especially by farmers and that water rights owners rarely sold any unused or surplus water rights, holding on to them as security in the event of future droughts or in the expectation of higher prices in the future.

Hearne (1995), also in a PhD dissertation, studied four areas in central and northern Chile, each characterised by an arid climate, water scarcity and a well-developed and commercially profitable agriculture sector. He found that with the exception of the Limari basin there was little trading in water rights in three of the four areas he studied. He considered that the reason for the limited market was related to the high costs involved in changing the rigid canal distribution to enable the transfer of water. Nevertheless, he suggested that there were economic benefits when water was transferred from low-value agricultural use to the higher-value urban supplies.

More questions in relation to the Water Code and why water rights markets have not been particularly dynamic appeared in a number of publications (Rios and Quiroz, 1995; Vergara, 1997a, 1997b, 1997c, 1998; Donoso, 1999; Dourojeanni and Jouravlev, 1999) Also, importantly, a discussion on social, environmental and institutional aspects had begun. The deficiencies of the Water Code identified by these authors echoed those of Bauer's, and included:

1   a lack of clear legal definition of water rights;
2   confusion with regard to return flows;
3   lack of hydrological information about water rights;
4   inadequate record keeping and registration of water rights;
5   inadequate infrastructure to allow inter-basin transfers;
6   over-abstraction of groundwater;
7   no provision to deal with environmental impact and third party effects arising from water rights trading;
8   in the case of mainly non-consumptive rights speculation, accumulation and hoarding, and excessive monopoly power.

Donoso (2003) in developing a case study analysis of water rights markets in Chile reported that the consensus of the response of participants during 'round table' discussions on the issue of whether the market permitted the reallocation of water rights from lower to higher economic values was that there was a lack of adequate information to answer the question formally. However, there was agreement on the following issues:

1   water companies were more active in Santiago – however, the amount transferred from one sector to another was almost equal to the historical transfer and proportional to the city's growth;
2   transfers from the agricultural sector to the basic sanitation sector related to those waters from agriculture that have been marginally used or that have fallen into disuse or covered by the expanding urban area;
3   there have been no transfers from intensive uses in agricultural activity unless land was sold or there was water surplus;
4   there have been transfers from agricultural activity in the Loa basin to mining companies, but in this area agriculture has not been significant.

In discussing groundwater exploitation, Donoso (2003) noted that although the Water Code provides for the regulation of the exploitation of groundwater by establishing protection, prohibition and restriction areas, aquifers have been exploited by several independent users with no clearly established water rights. Also that groundwater which has previously been used for irrigation to supplement surface water intermittently is now used for mining and water supply on a continuous basis, which has led to the overexploitation of aquifers.

Social impacts of the Water Code

The social impact of water rights trading on the indigenous population and small farmers has been left out of most studies. Water rights and water-use problems of poor farmers existed before the 1981 Water Code. Proponents of water markets do not consider that these groups have suffered as a result of water rights trading. However, Bauer (2004), based on interviews in the period 1992 to 2002 with staff of Chilean non-governmental organisations (NGOs) and several canal users associations indicated that the impact of the Water Code on poor farmers has been mainly negative. Romano and Leporati (2002) in a study of the distributive impact of the water market in Chile from 1981–1997, using the Limari basin as a case study, concluded that the distribution of water rights has worsened since 1981 and that the peasants' shares of water rights has significantly decreased as time has gone by, both in the aggregate and in per capita terms. Similar findings have been reported by others. Galaz (2004) reported a case in the Azapa valley, Arica, where farmers have lost their groundwater rights to a privatised water company. He notes that because peasant farmers are unable to mount legal challenges they accept violations of their water rights that benefit the more powerful water rights holders. Budds (2004) with reference to irrigation water rights transfers from peasant to commercial farmers in the groundwater dependent semi-arid La Ligua River basin in central northern Chile points out that by and large commercial farmers have the necessary awareness, financial resources and access to the legal system to acquire new water rights, which are lacking amongst peasant farmers. In a more recent paper, Budds (2009) criticised the results of a groundwater modelling study of the La Ligua basin, which enabled the wealthier and better educated farmers to secure additional groundwater rights (approximately, $1,550\,l\,s^{-1}$) at the expense of the peasants. The study, apparently, had failed to take into account data on the widespread (but illegal) use of groundwater by peasants which amounted to almost twice that of the actual legal abstractions. The same author in a paper on the water rights and indigenous groups in Atacama discusses the competing demands for water resources in the region from indigenous groups, mining, town supplies, wetlands and salt lakes (Budds, 2010). She suggests that the social outcome of the Water Code has been to favour stronger social actors and disadvantage weaker ones. The local Atacameños have been tempted by the large sums of money offered by mining companies and, since selling their water rights, many have migrated into towns. Some water rights have been bought back at market prices by local communities with the assistance of CONADI (National Corporation of Indigenous Development) while other water right acquisitions have been challenged. Similar findings have been reported by Madaleno (2007) with regard to the indigenous Aymara communities in the arid Tarapacá region of northern Chile.

In September 1993, the indigenous rights law (Ley indigína, 19.253) was promulgated with the object of protecting the rights of the indigenous (Aymara and Atacama) communities. The law considered that waters found in the areas of these communities are property to be owned and used by them, without, however, prejudice to the rights that third parties have registered under the 1981 Water

Code. New water rights, however, were not to be granted without guarantees to the normal water supply to the affected communities. The indigenous rights law was successfully invoked in 2009 by the Chilean Supreme Court in the case of the Aymara case against Agua Mineral Chuzmiza, a mineral water company which had obtained rights to use a sulphurous spring. The court upheld the water-use rights of the Ayamara community against the Chuzmiza water bottling company.

The 2005 revision of the Water Code

The shortcomings of Chilean water policy have been recognised by the government since the early 1990s and a number of discussion documents were prepared in response to this by the DGA. The latest document titled *Politica Nacional de Recursos Hidricos* (DGA, 1999) has been summarised by Bauer (2004). Its second half contains proposed reforms to the Water Code which include, *inter alia*: legal and institutional aspects, including fees for non-use of rights, reserving minimum ecological flows for new rights, integrated water management; protection of the environment; water resource use and water use; data collection and monitoring of water resources; and training and education of the public.

In recognition of these shortcomings, the Water Code was eventually reformed in 2005 (Modifica el Codigo de Aguas, Ley No 20.017, May 2005). In the reformed code the president was given authority to exclude water resources from economic competition in cases where doing so is necessary to protect the public interest. Other revisions included:

1   a requirement for the DGA to take into consideration ecological water flows and aquifer development when determining new water rights;
2   the introduction of licence fees for non-use as a deterrent against hoarding and speculation of unused water rights;
3   in order to limit the issue of water rights to genuine needs, a strengthening of the involvement of water users in public decisions and authorisation of the creation of water users associations;
4   recognition of the country's water communities as legal entities.

Summary

It had been hoped that the free transferability of water rights that was made possible by the 1981 Water Code would result in a lively market of many buyers and many sellers. The reality was different. The markets proved to be weak and generally dominated by large businesses. Other factors, geography and existing infrastructure among them, proved far more important in allocating water resources than the freedoms afforded to the market by legislation. The Water Code has not been kind to groundwater. Groundwater rights allocations have been generally optimistic, partly because of the difficulties in assessing groundwater resources, and also perhaps because of the natural desire of sellers to achieve a higher price. The Mapocho alluvial aquifer in Metropolitan Santiago is a good example of overallocation of groundwater rights. Fossil aquifers in the arid and semi-arid mining areas of northern Chile are also being overabstracted. Groundwater depletion led to the designation of aquifer protection zones and in the revision in 2005 of the 1981 Water Code. Interestingly, it took government intervention rather than the market to put in place measures to control overexploitation and to promote sustainability. A further consequence of the 1981 Water Code was the transfer of water rights from indigenous communities and small farmers in the north mainly to mining companies. The resulting socio-economic impacts led to the enactment of the 1993 indigenous rights law which was aimed at protecting the water rights of the indigenous communities. There is little doubt that the introduction of transferable water rights in Chile has brought about an appreciation of the value of scarce water resources and helped in the development of the country's major economic sectors, namely mining, and exports of high value agricultural products. However, this has been at the expense of adverse effects on water resources, the environment, indigenous communities and the poor.

Mexico

Water resources

The National Water Commission of Mexico (Comisión Nacional del Agua, abbreviated as CONAGUA, previously as CNA) in their publication *Statistics on water in Mexico*, 2010 edition (CONAGUA, 2010) compiled a comprehensive set of water statistics for the whole of the country and, unless otherwise indicated, much of the information in this section has been drawn from this publication. In addition, data on groundwater abstractions, recharge and water right concession titles were obtained from Registro Público de Derechos de Agua on the CONAGUA website.

In the last 60 years the population of Mexico has quadrupled from just below 26 million in 1950 to approximately 103 million in 2005 and 112 million in 2010. There has been a movement of population from the rural areas to the cities. In 1950, 57.4 per cent of the population lived in the rural areas, but by 2005 only 23.5 per cent. The concentration of people in urban areas has placed a strain on water resources and the environment.

There are two main mountain ranges running north–south, namely, the Sierra Madre Oriental along the Gulf of Mexico coast to the east and the Sierra Madre Occidental along the Gulf of California and the Pacific Ocean in the west. Between them is the Central Plateau. Mexico has a variable climate, generally arid to semi-arid in the north and centre, hot and humid in the south and southeast and temperate and humid in the mountainous areas. The mean annual rainfall for the whole country for the period 1941–2005 was 773.5 mm and a little less, 759.6 mm, for the period 1941–2000. Most of the rain occurs between June and September. Rainfall generally increases southwards. In the northwest, the average annual rainfall is 100–250 mm, in the centre 250–500 mm, in the south and southeast 1,000 mm to more than 2,000 mm, and in the Sierra Madre Oriental 700–1,500 mm. Water requirements are greatest in the drier north and centre of

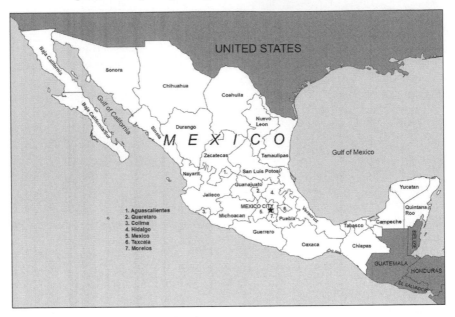

Figure 6.2 Administrative map of Mexico

the country, where the main economic activity and population growth are taking place. Irrigated agriculture is the main water user. In 1950, agriculture, together with forestry and fishing, accounted for 55 per cent of the gross domestic product (GDP) but in 2008 their contribution was less than 4 per cent. Agriculture is, however, the main source of income for a large number of poor farmers and rural labourers (World Bank, 2009).

From 1997 Mexico has been divided into 13 hydrological-administrative regions, which are usually denoted in Roman numerals (I to XIII) from northwest to south-southeast. The total amount of annual renewable water resources of Mexico is 459,351 Mm³, comprising 378,530 Mm³ measured surface water runoff and 80,821 Mm³ recharge to aquifers. Of this total amount, approximately 69 per cent is found in the south and southeast where less than a quarter of the population lives. In the greater metropolitan Mexico City area (Waters of the Valley of Mexico) where approximately 20 per cent of the people live, annual renewable resources are less than 1 per cent of the nation's total water resources. Similarly, approximately 61 per cent of aquifer recharge occurs in the south and east leaving less than 31,000 Mm³a⁻¹ for the arid and semi-arid northwest and centre of the country.

There are 653 aquifers in Mexico, 84 in the south and east and 569 in the northwest and centre. The total annual groundwater abstraction in 2008 was approximately 29,500 Mm³ and in 2011 approximately 29,785 Mm³, or approximately 36.5 to 36.9 per cent of the total aquifer recharge. However, as in the case of Chile this is misleading. It is only in the wet south and east that abstraction does not exceed recharge. In the arid and semi-arid northwest and centre aquifers have been increasingly overexploited and a few have suffered seawater intrusion. In 1975, there were 32 overexploited aquifers, in 1980 there were 80, in 2002 the number has risen to 97 and in 2008 to 101. In the Mexico City metropolitan area aquifer recharge is about 44 per cent of the annual abstraction of approximately 1,419 Mm^3 and as a result groundwater levels have been falling by about 1 m a^{-1}. In addition there has been land subsidence at 0.1–0.4 m a^{-1} (Tortajada, 2006). Nationally, groundwater levels have been falling between 1 and 5 m a^{-1}. Aquifers are overdrawn by approximately 6,500 Mm^3a^{-1} (CONAGUA, 2011). Data from the *Diario Oficial de la Federación* (2009), Volume DCLXXI, No 20 suggest that the groundwater deficit in 2002 from 282 aquifers was 930 Mm^3 of which 67 per cent was from the arid Baja California peninsula and the northwest.

In 2008, approximately 77 per cent of the total water use of 79,800 $Mm^3 a^{-1}$ (comprising 50,200 $Mm^3 a^{-1}$ surface water and 29,500 $Mm^3 a^{-1}$ groundwater) went to agriculture, 14 per cent to public supply and 9 per cent to industry, agricultural industry and thermal power. Groundwater use in agriculture was approximately 69 per cent of the total, public water supply 24 per cent and industry 7 per cent. The share of groundwater for irrigation has increased over the years as indicated by the number of agricultural wells which has increased from around 10,000 in 1970 to more than 95,000 in 2000 (World Bank, 2009). In the period 2001–2008, agricultural use of groundwater increased by approximately 21 per cent, public supply use by 11 per cent and industrial use by 36 per cent.

Water legislation

The National Water Law (Ley de Aguas Nacionales) (NWL) was promulgated in 1992. It affirmed the ownership by the nation of surface water and groundwater and appointed CONAGUA, which was created in 1989 to develop and implement the water policy and transition process, as the only federal water authority in the country to regulate water, including the granting of concession user titles or deeds. Although, under the new law, CONAGUA was formally given a largely policymaking and oversight role, in practice it maintained a great deal of influence over operations and retained key strategic functions within its jurisdiction (Wilder and Romero Lankao, 2006). The state retains strong administrative powers. The 2004 NWL reform stressed the efficient use of water and the recovery of charges even from agriculture. CONAGUA may limit abstractions to protect the environment. By 2009, 145 groundwater protection zones were established restricting groundwater extraction. These covered much of the country, including the northwest, the centre and parts of the south and east.

Changes in the Agrarian Law in 1992 and the NWL strengthened the private water rights of individuals, and thus of the *ejido* farmers, over those of the community. *Ejidos* are parcels of communal land, which were created during the land reforms following the 1917 Mexican revolution, partly in order to ensure a more equitable distribution and use of land, which at the time was largely privately owned, and partly to help with the livelihoods of peasant farmers. The rights to *ejido* property by farmers were purely usufructuary, and *ejido* lands could not be sold or leased. The 1992 amendment of Article 27 of the Agrarian Law changed this and established a process by which *ejido* members could own their parcel of ejido property. Land rentals were freed completely and sales of land, not allowed before, could take place within the *ejido* but not with outsiders, unless the *ejido* member had acquired property rights for his parcel of land.

The NWL defines water rights as volumetric concessions specifying the type of use. They are granted for periods of 5 to 50 years, although the second set of Presidential Decrees of October 1996 limited the duration to 10 years. In order to avoid hoarding and monopolies, water rights are cancelled if not used for three consecutive years. Groundwater rights may be transferred separate from land in areas approved by CONAGUA. Regarding volumes, they were initially based on availability, third-party effects and a requirement for the user to demonstrate effective use. The second set of Presidential Decrees simplified procedures only requiring a declaration under oath by the user and verification by CONAGUA by checking a statistically representative sample. Users are free to trade their water rights within irrigation districts, without the intervention of CONAGUA. Users are also free to trade their rights when only the user changes or in areas designated by CONAGUA (regulated aquifers), although in these cases all transactions must be registered. All other transactions have to be approved by CONAGUA in order to protect the environment and third parties. Dissatisfied users can apply to CONAGUA or to the law courts. The State Engineer may act as an arbitrator in disputes between stakeholders.

There was no consultation with stakeholders prior to the promulgation of the 1992 NWL. However, during its implementation there were vigorous campaigns encouraging people to register their water rights by pointing to the benefits of private ownership, security and tradability of water rights. Formal procedures in the registration of water rights improved transparency.

Water rights transfers

One of the important outcomes of the NWL has been the registration of water rights. Before the enactment of the NWL there were only 2,000 registered water concessions in the whole country. Initially the registration process was slow, but, as a result of simplification of the procedures and intense campaigning, the number of registered concessions rose to approximately 320,000 by 2000 (Garduño, 2005). In 2008, the number of registered concessions was 360,301, comprising 242,953 groundwater registrations and 117,348 surface water registrations. In 2011 the total number of registrations was 370,333 of which 251,956 were for groundwater and 118,377 for surface water. The total authorised abstraction volume in 2008 was $79,752 \text{ Mm}^3 \text{ a}^{-1}$ (CONAGUA, 2010) which is similar to the quoted volume of $79,800 \text{ Mm}^3 \text{ a}^{-1}$ for total water use in the same publication. On this basis, the authorised abstraction for groundwater is $29,500 \text{ Mm}^3 \text{ a}^{-1}$.

Temporary water transactions were common within irrigation districts before the 1992 NWL. Authorised transactions between 1992 and 1998 were few: 573 in groundwater and 7 in surface water (Garduño, 2005). The number of transactions not requiring CONAGUA approval is not known. Also, Kloezen (1998), though admittedly some time ago, noted that empirical evidence of water trades in Mexico was scarce. He reported on water right trades in the Alto Rio Lerma Irrigation District, in the State of Guanajuato, central Mexico, which has a command area of 112,772 hectares, and is served by four major dams with a capacity of 2,140 Mm^3 (although, the volume available for irrigation in 1982–1996 was only 880 Mm^3) and approximately 1,700 tubewells. There were 11 water user associations (WUAs) in the district, each having a proportional allocation of available storage which dates back to many years before the NWL came into being. Water trading of concessions is of two types, trading for the entire period of the concession and trading or renting of a fraction of the concession for a period much shorter than the duration of the concession. Most trades take place during the summer irrigation season, with some WUAs depending almost entirely on bought water. Volumes bought in 1995 amounted to 22 Mm^3, in 1996 to 8 Mm^3 and in 1997 to approximately 10 Mm^3. As a percentage of the total volume of water used summer trades ranged between 2 per cent and 86 per cent. If the two extreme values are excluded the average water trade was about 21 per cent of total use. However, the share of traded water to all water used in winter and summer was less than 5 per cent. Moreover, in terms of the total volume of 880 Mm^3 of surface water available to the district for irrigation, the volume of trades was less than 2.5 per cent. In 1996–1997 the traded volume of 10 Mm^3 corresponded to approximately 1.5 per cent of the allocated volume of 640 Mm^3 to the district by CONAGUA. Trading was within the agricultural sector between the WUAs and it is not clear whether it included groundwater trades. Kloezen concluded that the NWL through the development of WUAs has encouraged bulk trade between the WUAs and a successful system of reallocating parts of their yearly water concessions through a water market. However, transaction costs remained low because trading has been local between the WUAs and did not involve transport costs or externalities on other water users. Also, WUAs place social and political returns higher than economic returns. By helping other WUAs with access to additional water they ensure the performance, sustainability and credibility of all WUAs. The experience of the Alto Rio Lerma Irrigation District suggests that marketing helps in re-allocating water between users. However, as the allocation of water volumes is on the basis of hydrological factors, trading has no significant role to play in the management of water resources. Moreover, the volume of trades in comparison to total water use is too small to have any serious impact on water resources management.

Fortis and Ahlers (1999a, 1999b) analysed transactions by individual users in Distrito de Riego (DR) 017 – Comarca Lagunera, Coahuila and Durango. They describe an active market, characterised by many temporary transactions, especially in the *ejido* sector. This leasing practice is part of the overall economic strategy of *ejidatarios* and other small farmers who have experienced very difficult situations since the 1990s due to the transformation of the agricultural sector and the state's role. In addition, in the mid-1990s a severe drought in the area increased pressures on the water system. These researchers reached a number of conclusions. First, although the water market was quite dynamic, there were many small sellers but only a few large buyers, and the latter appear to have taken advantage of their political influence. Second, there was uncertainty with regard to water rights in the medium and long term, although this was more due to the economic and political instability of the agricultural sector than to water legislation. And third, price signals were rather confusing because government continued to charge low prices for irrigation water, apart from shortage variations caused by droughts or floods.

Asad and Garduño (2005) in their bibliographic review of empirical studies suggested that the basic conclusions drawn by the various researchers were quite similar. First, the state with its strong regulatory power seems to be the controlling factor in the operation of different local water markets, and although it may be justifiable in many cases, it has weakened the legal security of water use rights. Second, where there have been dynamic markets the analysis is complicated because many transactions do not involve entitlements and/or legal owners or seem to be the result not of economic factors but to stem from social and political reasons. And, third, the legal definition of individual entitlements within irrigation districts and WUAs continues to be incomplete or inexact in many cases. There has been generally an over-concession of water use rights, which has led to social and economic problems. This has been worse for groundwater because in a large number of aquifers extraction for many years has been from storage. The application of various instruments, including regulatory, participatory and economic, all initiated by CONAGUA, has achieved significant progress in addressing these problems but progress has not been long lasting.

Levina (2006) indicated that 510 water trades were registered (no date given) equivalent to a volume of 143.27 $Mm^3 a^{-1}$ in the water rights public registry (REPDA, Registro Público de Derechos de Agua), mainly in dry regions or areas with tight water balances; 222 transfers (47 $Mm^3 a^{-1}$) were between irrigation users and 40 transfers (61 $Mm^3 a^{-1}$) between industrial users.

There has been a general lack of empirical studies of water markets in Mexico in the last 10 years. Even during the 1990s, when much was written on the subject, quantitative information was generally poor and with regard to groundwater very scanty. The comprehensive publication by CONAGUA (2010) on national water statistics does not include data on water transfers or water rights trading. Also, interestingly, water rights trading is not on the list of initiatives in the 2030 Water Agenda to achieve a balance in supply and demand (CONAGUA, 2011). The initiatives focus on:

1   the strengthening of water resources management instruments;
2   a more prominent role of aquifer management by the technical groundwater committees (*comités técnicos de aguas subterráneas* or COTAS);
3   more involvement of irrigation user associations and COTAS in the effort to save water and in the application of irrigation technology; and
4   the strengthening of systems to measure and ensure compliance with the volumes allocated and authorised in an agricultural year.

COTAS

COTAS were developed in the 1990s by CONAGUA in an effort to control overabstraction of groundwater. They are consultative committees, without legal status or decision-making powers, comprising aquifer users, government water agencies and organised groups from civil society (Wester *et al.*, 2011). The idea was that stakeholders interacting together could achieve reductions in groundwater abstractions voluntarily, thus avoiding the long and expensive legal process of reducing concessioned volumes. In a detailed assessment of the impact of COTAS in Guanajuato State, Central Mexico, Wester *et al.* (2011) considered that the COTAS did not achieve reductions in groundwater abstraction, as had been hoped for. In fact abstractions marginally increased, while groundwater levels continued to decline. Participation by users was very small, 3–10 per cent. Industrial users and potable water suppliers argued that the agricultural sector, which uses more than 80 per cent of the groundwater, was largely responsible for the overexploitation; and even if it were to reduce abstractions substantially, the impact on groundwater levels would be minor. Farmers, on their side, were largely sceptical and feared that they might lose control and their freedom over groundwater pumping. The authors concluded that the Guanajuato experience suggested that the COTAS approach of self regulation was not sufficient to ensure groundwater sustainability without the enforcement of water legislation, especially with regard to well permits and pumped volumes.

Water banks

Water banks in Mexico have been set up by the government with the aim of regulating and facilitating the acquisition and transfer of water rights. Other aims include preventing speculation and hoarding, and avoiding the development of informal water markets. In their role as facilitators of water rights transfers, water banks are to provide advice and guidance to users and verify that the water use rights to be transferred comply with the applicable regulations (CONAGUA/OECD/IMTA, 2010). The first water bank was set up in 2008. By the end of 2009 there were six operating water banks and by March 2012, 30 (CONAGUA website Bancos del Agua en Mexico). The government hope that water banks will help in the reallocation of water rights to other uses and in the sustainable management of water resources, especially in the northern and central regions where water resources are scarce. It is perhaps too early to assess the contribution of water banks in water resources management and the control of groundwater abstraction, and whether they will enable the transfer of water rights across different sectors in large volumes. Probably their role will be useful mainly during drought periods in relation to temporary transfers.

Summary

From the foregoing, it seems that, so far, water rights trading in Mexico has been limited mainly to surface water between farmers within irrigation zones based on quotas allocated by the government. There is no evidence of significant movement of groundwater from irrigation to higher value urban or industrial use. Government policy whilst not discouraging water rights trading appears to have placed priority on the formal registration of water rights, which is an important first step in ensuring the security of water rights. Regarding the controlling of overabstraction, particularly of groundwater, non-market instruments, such as prohibition measures and protection zones, have been applied. As in Chile, water rights in Mexico have been oversubscribed and allocated volumes exceed available quantities. Indeed, Asad and Garduño (2005) have come to the conclusion that a Water Rights Adjustment Programme (WRAP) is required involving government intervention to cancel water rights by offering compensation to farmers, for groundwater 3,600 m^3 per year at cost of 6.8 billion pesos (approximately US$ 0.62 billion), and for surface water 1,500 Mm^3 per year at cost of 2.6 billion pesos (approximately US$ 0.24 billion).

The establishment in the last few years of government regulated water banks to facilitate the transfer process seems to be Mexico's approach to water rights trading. The initiatives put forward in Agenda 2030 for achieving a balance between water supply and demand rely on government intervention and investment, which in some way illustrates Mexico's view on the role of the free market in water resources management.

Western United States of America

Water resources

The climate of the western USA is generally arid to semi-arid. Coastal California has a Mediterranean climate and southeastern Texas (a southern state) along the Gulf of Mexico is hot and humid. Annual rainfall is generally 250–500 mm. In southeast California, rainfall is less than 125 mm increasing to 1,000–2,000 mm in the coastal range and Sierra Nevada mountains in northern California, and in southeast Texas, 600–1,000 mm (USGS n.d.). There are three main river systems: the Rio Grande, which runs through New Mexico and Texas and discharges into the Gulf of Mexico; the Colorado which runs through Colorado, Utah and Arizona and discharges into the Gulf of California; and the Sacramento and San Joachim Rivers, both in California, discharging into the San Francisco Bay on the Pacific Ocean. The Colorado River and Rio Grande have their headwaters in the Rocky Mountains. The Sacramento River runs southwards and drains the Klamath Mountains and Cascades Range in the north. The San Joachim River runs northwards and drains the Sierra Nevada mountains in the east.

The principal aquifers in the western USA are: the High Plains (or Ogallala) aquifer, the Central Valley aquifer system, the Basin and Range basin-fill aquifers and the Californian Coastal Basin aquifers (Maupin and Barber, 2005). Table 6.1 summarises the exploitation of the principal aquifers in 2000. In all aquifers, irrigation abstraction predominates: 97 per cent in the High Plains, 91 per cent in the Central Valley of California and 81 per cent in the Basin and Range basin-

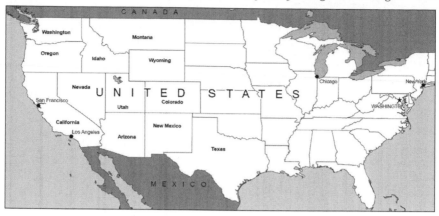

Figure 6.3 Administrative map of the USA

fill In the Coastal Basins of California, abstraction for irrigation was 51 per cent; public supply was the other large user at 46 per cent. Industrial use is generally less than 1 per cent. Nebraska, Texas and Kansas use about 88 per cent of the groundwater from the High Plains aquifer. Arizona uses more than half of the groundwater of the Basin and Range basin-fill aquifers.

In 2005, the total fresh groundwater abstraction in seven western states (California, Nevada, Utah, Colorado, Arizona, New Mexico and Texas) was approximately 38,400 Mm³, of which approximately 38 per cent was in

Table 6.1 Groundwater exploitation of the principal aquifers in the western USA (based on Maupin and Barber, 2000)

Aquifer system	Extent	Area (approximate) (km²)	Groundwater abstraction 2000 (approximate) (Mm³)			
			Irrigation	Public supply	Industry	Total
High Plains (Ogallala)	Colorado, Kansas, Nebraska, New Mexico, Oklahoma, South Dakota, Texas and Wyoming	456,000	23,500	600	100	24,200
Central Valley	California	52,000	21,300	2,000	200	23,500
Basin and Range basin-fill	Mostly in Nevada, parts of eastern and southern California, eastern Utah and southern Arizona		6,280	1,400	80	7,760
Coastal Basin	California		2,430	2,170	150	4,750

California and 29 per cent in Texas (Kenny *et al.*, 2009). Groundwater abstraction represented approximately 32 per cent of total (surface water and groundwater) abstraction. On average, approximately 78 per cent of the groundwater was used for irrigation. Colorado had the highest irrigation use at approximately 92 per cent, followed by California and Texas at 81 per cent and 78 per cent, respectively. Groundwater exploitation in the whole of the USA has progressively increased from approximately 47,000 Mm^3 in 1950 to 113,000 Mm^3 in 1975, and since then, has varied in the range 101,000 Mm^3 and 116,000 Mm^3.

The result of the intensive exploitation of groundwater in the western USA has been a general depletion in storage and there have been other impacts, such as land subsidence in the California Central Valley and a reduction in springflows and discharges into wetlands (see, Edwards aquifer in Texas, below). In the High Plains aquifer groundwater level decline between predevelopment (about 1950) and 2005 ranged between 26 m and 84 m, and was on average 4 m (McGuire, 2007). The largest falls were in Texas (more than 45 m). McGuire (2007) estimated the reduction in aquifer storage since predevelopment to be approximately 312,000 Mm^3 or 9 per cent of the total aquifer storage. The Central Valley aquifer in California was recently the subject of a comprehensive study and modelling (Faunt, 2009). About one-sixth of irrigated land and about one-fifth of groundwater use in the USA are in the Central Valley. It is one of the most productive agricultural regions in the world, producing 25 per cent of US food, including cereal grains, hay, cotton, tomatoes, vegetables, citrus, nuts and grapes. The estimated value of agricultural produce in 2002 was US\$ 17 billion. Irrigated agriculture has been supported by surface water through an extensive system of dam reservoirs and canals diverting water from the main river systems (California Central Valley Project and State Water Project) and by groundwater. Pressures on groundwater resources have been recently increasing, partly because of expanding urbanisation – between 1970 and 2000 the population in the Central Valley doubled, reaching 6.5 million in 2005 – and partly because of environmental pressures requiring water allocations to maintain the integrity of the ecosystems. The Central Valley aquifer is recharged by rainfall infiltration, irrigation returns and river leakage. Irrigation returns constitute a substantial proportion of recharge. For the period 1962–2003, the average groundwater recharge was approximately 11,225 $Mm^3 a^{-1}$ and abstraction 12,950 $Mm^3 a^{-1}$, leaving a deficit of 1,725 $Mm^3 a^{-1}$. During dry years the deficit increases to 11,100 $Mm^3 a^{-1}$ but in wet years more than 15,400 $Mm^3 a^{-1}$ is taken into storage (Faunt, 2009). Overall there has been a steady decrease in groundwater storage, which over the 40-year period (1962–2003) amounted to almost 70,000 Mm^3. The result of excessive groundwater pumping has been a decline in water levels, particularly in the southern part of the San Joachim Valley, where by 1961 water levels had fallen between approximately 35 m and 90 m, and in places by more than 120 m. After 1960, water levels recovered due to a reduction in groundwater abstraction and the importation of surface water. Interestingly, during the drought periods of 1997 and 1987–1992 when surface water imports diminished, increases in groundwater abstraction depressed water levels to their pre-1960 levels. Groundwater withdrawals have also caused land subsidence, particularly in the San Joachim Valley.

Water legislation

The main features of groundwater rights in western USA are summarised in Table 6.2. Prior appropriation is the basic groundwater allocation doctrine in most western states, although many states have modified it in response to overabstraction or other circumstances. Texas still applies the absolute ownership doctrine and Arizona the reasonable use doctrine. In California both the prior appropriation and correlative rights doctrines apply. Beneficial use defines the measurement and limit of use in most states. Where a permit system has been introduced, the volume of water that can be extracted in a given period is generally stipulated. There is no restriction on the duration of the water right, but it can be lost if not used. Water rights are transferable independent of land ownership, except under the reasonable use doctrine. Disputes are resolved in ordinary courts. Specialist water courts exist in some states, such as Colorado. In New Mexico, the state engineer may act as an arbitrator. There is generally no specific provision in groundwater legislation for the environment. However, the state may appropriate waters or withdraw them from appropriation to protect stream flows or lake levels for fish and wildlife, or by invoking the federal reserve rights or the federal Wild and Scenic Rivers Act. The federal and Indian Reserved Rights doctrine protects Indian water rights irrespective of whether or not the water had been used before the arrival of European settlers. Historically, the European settlers and the state and federal governments appropriated water resources without any consultation with the indigenous Indian populations. The Indian Reserved Rights doctrine only protects those native Americans who have a treaty with the United States, but does not apply to those tribes or communities which do not have such treaties. Today there is strong stakeholder participation representing the interests of communities and organisations, but there is no requirement for this in water legislation. Transparency has improved as a result of formal procedures in the allocation of water rights in permit systems.

Water rights transfers

The western USA, and particularly California, has had a long history of water rights trading, both formally and informally. The most established market for federal waters operates in the Northern Colorado Water Conservancy District. Here annual entitlements are freely transferable and about 30 per cent of the water delivered to the district each year passes through the rental market. During the late 1970s and 1980s, there were approximately 6,000 transactions in Utah, 1,455 in New Mexico, and 1,500 in Colorado (Steinhart, 1990). In Arizona, after 1990 when groundwater was made freely transferable, the cities of Phoenix, Tucson, Mesa and Scottsdale acquired more than 50,000 acres of farmland in order to retire the fields and to utilise the water.

Czetwertynski (2002) in a survey of water rights transactions in 14 western states (Arizona, California, Colorado, Idaho, Kansas, Nevada, New Mexico, Montana, Oklahoma, Oregon, Texas, Utah, Washington, Wyoming) identified

Table 6.2 Summary of the main features of groundwater rights in the western USA (partly based on Bryner and Purcell, 2003)

State	Allocation doctrine	Volumetric definition of right	Duration of right / loss of right
Arizona	Reasonable use.	Beneficial use is the basis, measure and limit of the use of water.	Forfeiture of right after five years of non-use.
California	Prior appropriation: applies to groundwater users who do not own the overlying land. Correlative rights doctrine applies to users who own the overlying land. Correlative rights are superior to appropriative rights, the latter being limited to the surplus.	Limited to beneficial use without waste.	No limitation.
Colorado	Prior appropriation. Modified in order to increase economic development (regulate groundwater pumping and integrate surface water appropriations with groundwater appropriations).	In designated basins, permits stipulate the maximum annual volume of the appropriation in acres per year and the maximum pumping rate in gallons per minute. In non-designated basins, the permitted annual withdrawal is based on an aquifer life of 100 years and is not to exceed 1 per cent of the total amount of groundwater, exclusive of artificial recharge, recoverable from a specific aquifer beneath the overlying land.	No limitation.
Idaho	Prior appropriation; however, groundwater appropriations, other than those related to domestic wells, require a permit and licence.	Beneficial use. In critical groundwater areas, the director of the Department of Water Resources may order right holders to cease or reduce groundwater withdrawals on a time priority basis until he determines there is sufficient groundwater.	Water right is lost and forfeited if the permit holder fails in a period of five years to put the appropriated water to beneficial use.
Montana	Prior appropriation. Permits required to appropriate water. Depending on the quantity of water required different criteria apply. Groundwater appropriations may be limited by declaring control areas. Also, by legislative means, although these have been used mainly for surface water.	Based on beneficial use.	Water right is lost if not used for 10 successive years.

State	Allocation doctrine	Volumetric definition of right	Duration of right/loss of right
New Mexico	Prior appropriation. Withdrawals require permits issued by the state engineer.	Beneficial use is the basis, the measure and the limit of the right to water use.	No limitation.
Nevada	Prior appropriation. In designated basins (these are basins having reasonably ascertainable boundaries) permits are required to appropriate water, which are issued by the state engineer. No permits required for deep wells, greater than 2,500 feet.	Beneficial use is the basis, the measure and the limit of the right to water use. Permits relate to a specific volume of water and conditions preventing adverse effects on other users.	No limitation.
Oregon	A permit is required to use groundwater or to construct a well. The Water Resources Commission may designate critical groundwater areas where corrective measures are taken, including no further groundwater appropriation and the establishment of permissible abstraction rates.	Beneficial use without waste is the basis, measure and extent of the right to appropriate water.	Loss of water right, if right owner ceases to use all or part of the right for five consecutive years.
Texas	Absolute ownership doctrine (rule of capture rule). Groundwater Conservation Districts created by Texas legislature to conserve, preserve, protect, recharge and prevent waste of groundwater resources. Edwards Aquifer Act 1993 to protect the diminishing springflows and ecology. The Act introduced a permit system (see discussion on the Edwards Aquifer below) prohibiting any person from withdrawing groundwater in the district unless authorised by a permit, each permit specifying the maximum rate and the total annual extraction volume. However, the ruling of the Texas Supreme Court of February 2012 reaffirmed the absolute ownership doctrine (see, Chapter 3).	No limitation providing: (a) non-wasteful use; (b) no malicious injury on neighbour; (c) not knowingly causing subsidence on neighbours land; (d) no deviated ('slant') wells reaching aquifer beneath neighbour's land.	No limitation.

Continued...

Table 6.2 continued

State	Allocation doctrine	Volumetric definition of right	Duration of right/loss of right
Utah	Prior appropriation but users require a permit issued by the state engineer.	Beneficial use is the basis, the measure and the limit of the right to the use of water.	Issued by state engineer for a limited amount of time.
Washington	Prior appropriation under a permit system.	Permit states the amount of water allowed and the beneficial use to which it may be applied. The Department of Ecology may impose limits on groundwater withdrawals in designated areas to prevent overdraft.	Loss of right, if water is not beneficially used for a period of five successive years.
Wyoming	Prior appropriation under a permit system. The State Board may designate control areas where, inter alia: the use of groundwater is approaching the current recharge rate or where groundwater levels are declining or have declined excessively. Measures in control areas may include: no further appropriations, apportioning permissible withdrawals among appropriators, the ceasing or reducing withdrawals from junior appropriators.	In control areas: quantity to be beneficially used in gallons per minute and acre feet per year. Proposed appropriations within 15 miles of Yellowstone Park require study to show that hydrothermal springs are not affected. Outside control areas, permits for beneficial use issued as a matter of course, unless contrary to the public interest.	No limitation.

1,068 sales and 552 leases. Of the 1,620 transactions recorded only 8.6 per cent related to groundwater. Also, of the 1,068 sales, 738 or approximately 70 per cent related to the Colorado-Big Thompson scheme. Of a total of 19,362 Mm³ recorded transfers in the 14 western states only a small minority, approximately 3.5 per cent, involved permanent sales. The great majority of transfers were in the form of leases. Also in all states, farmers were very seldom the buyers of water rights, the main buyers being municipal and industrial water providers.

According to Hanak (2003), the water market in California in 2001 represented only 3 per cent of all water used for municipal, industrial and agricultural purposes. Water transfers have been increasing steadily from about 60 Mm³ in 1985 to more than about 1,500 Mm³ in 2000 and 2001. The market grew as a result of direct intervention of the Department of Water Resources making dry-year purchases to offset lower deliveries to state water projects and wildlife refuges. Thus the large increase to about 1,400 Mm³ in the first year of the drought in 1991 was in response to state-sponsored spot markets whilst about 50 per cent of the increases since 1996, a period when hydrological conditions were favourable, have been mainly due to environmental demands imposed by environmental regulations. There has been considerable resistance to water trading by communities in the source regions, due to effects on the local economy and on local groundwater users (Howitt and Sunding, 2004). Regarding the first, the fallowing of land in the 1990s by the state due to water purchases for the 1991 drought water bank led to a slightly reduced demand for labour and other farm inputs, and a decrease in the supply of raw materials to local processors. Losses in county income ranged between 3.2 per cent in Solano County, where 8 per cent of the acreage was fallowed for transfers, to 5 per cent in Yolo County, where 13 per cent of the irrigated acreage was fallowed (Howitt, 1994). Farmers who sold their surface water allocations replaced them by pumping additional groundwater. This led to accusations of reducing the quantity and quality of groundwater to other users, and concerns that groundwater was being mined. Local resistance to water trading has taken the form of local ordinances that mandate the acquisition of a permit before exporting groundwater or abstracting groundwater to substitute for exported surface water. Since 1996 this has resulted in a reduction in groundwater exports of 1,150 Mm³ or 19 per cent and of total sales of 970 Mm³ or 14 per cent. (Hanak, 2003). When crop prices are low, farmers conserve water for the water market by not cultivating their land.

Water rights transfers for the period 1987–2009

A water transfer database has been prepared for 12 western states (Arizona, California, Colorado, Idaho, Montana, New Mexico, Nevada, Oregon, Texas, Utah, Washington and Wyoming) for the period 1987–2009 by the Bren School of Environmental Science and Management of the University of California funded by the National Science Foundation and the California Water Resources Centre. The database is presented in the form of an Excel spread sheet and may be found at the Bren website (www.bren.ucsb.edu/news/water_transfers.htm). It contains data on the number of transactions, volumes transferred across different sectors and the prices paid. The database classifies transferred water quantities in terms of annual flows and committed flows. The annual flow is the amount that is transferred in the first year of the transaction, which is further subdivided into minimum, maximum and average. The committed flow reflects the long-term transfer quantity for water rights transferred for long periods or in perpetuity. It is determined by discounting the flow of water over time into the year the water was transferred, using a discount rate of 5 per cent.

The number of transactions for the 12 states for the period 1987–2009 is shown graphically in Figure 6.4. There is a general upward trend, although in the last few years the number of transactions in some states has decreased. In California transactions between 1997 and 2006 remained more or less the same, the increase in 2007–2008 was mainly due to environmental transfers. Transactions tend to be higher in the drier periods, such as, during the drought of the early 1990s.

A summary of the total number of transactions in the period 1987–2009 and the average number per year is presented in Table 6.3. The number of transactions for most states has remained generally small, on average between three and ten per year. In Texas and California the number of transactions is a little higher, 15 and 30 per year, respectively. Colorado, due to water from the Colorado-BigThompson scheme, predominates with an average of 97 transactions per year or approximately 51 per cent of all transactions. The database does not generally differentiate between surface water and groundwater transactions, but the great majority appear to relate to surface water. About three-quarters of all transactions involve short-term leases of one year or less; long-term leases of 50 years or more are only a small percentage.

In the 23-year period 1987–2009, the average annual (committed) volume transferred in the 12 states was 199,424 Mm^3 or approximately 8,671 Mm^3 a^{-1}. In comparison with the 2005 total water withdrawal from the 12 states of 184, 950 Mm^3 (Kenny *et al.*, 2009), the proportion of transferred water is very small, amounting to approximately 4.7 per cent. The lowest is in Montana, 0.3 per cent, and the highest in Arizona, approximately 14 per cent. In California, Colorado and Texas, transfers range between 5 and 6 per cent of the total abstraction (Table 6.4).

Transaction volumes appear to be related to dry or wet periods. For example, committed transfer volumes were lowest during the drought of the early 1990s, followed by an increase after 1996, and a decrease after 2005 (Figure 6.5). Annual transfer volumes remained at more or less the same level during the drought (Figure 6.6).

Analyses of water transfers by Brewer *et al.* (2008) and Donohue (2009) for the period 1987–2007 indicated that the majority of trades (79 per cent) originated from the agricultural sector. Transfers from the urban sector amounted to 16 per cent and transfers that have more than one original use, 5 per cent. Transferred volumes from agriculture amounted to 54 per cent of committed flows and 62 per cent of annual flows and from the urban sector to 28 per cent of committed flows and 22 per cent of annual flows. Transfers of committed flows within the

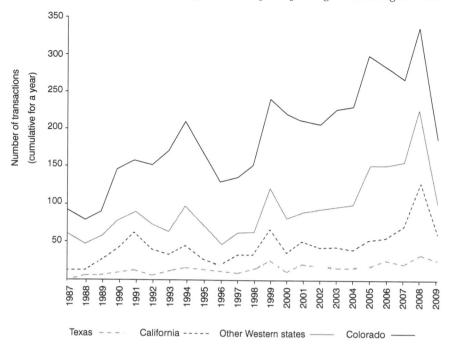

Figure 6.4 Water rights transactions in western USA for the period 1987–2009

Table 6.3 Total and annual average number of water rights transactions in western USA for the period 1987–2009 (dataset www.bren.ucsb.edu/news/water_transfers.htm)

State	Total number of transactions (1987–2009)	Annual average number of transactions (round figures)	Per cent of total number of transactions (per cent)
Arizona	238	10	5.4
California	692	30	15.7
Colorado	2,228	97	50.5
Idaho	148	6	3.4
Montana	59	3	1.3
New Mexico	153	7	3.5
Nevada	201	9	4.6
Oregon	130	6	2.9
Texas	346	15	7.9
Utah	87	4	2.0
Washington	59	6	1.3
Wyoming	66	3	1.5
Total/Average	4,407	192	100

Table 6.4 Transferred (average annual committed) water volumes in comparison to total withdrawals in western USA (1987–2009) (dataset www.bren.ucsb.edu/news/water_transfers.htm)

State	Total withdrawal 2005 (surface water and groundwater) (Mm³)	Average annual volume (Mm³) (committed) transferred (1987–2009)	Per cent volume transferred (per cent)
Arizona	8,634.37	1,247.7	14.4
California	45,392.1	2,340.6	5.1
Colorado	18,872.3	927.1	4.9
Idaho	27,013.2	579.3	2.1
Montana	13,938.3	39.6	0.3
Nevada	3,293.4	140.0	4.2
New Mexico	4,613.2	263.2	5.7
Oregon	9,978.9	523.7	5.2
Texas	32,687.3	2,059.4	6.3
Utah	6,673.1	270.4	4.0
Washington	7,746.3	216.7	2.8
Wyoming	6,105.7	62.9	1.0
Total/average	184,948.17	8,670.6	4.7

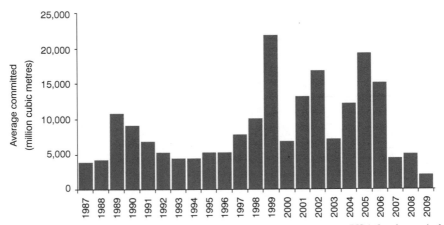

Figure 6.5 Average committed water rights transfer volumes in western USA for the period 1987–2009

agricultural sector itself were 10 per cent, from agriculture to urban supply 30 per cent and from agriculture to the environment 14 per cent.

Water rights transfer prices reported by Brewer *et al.* (2008) and Donohue (2009) for the period 1987–2007 are shown in Table 6.5. As would be expected, sales of water rights, which involve the transfer of a quantity of water in perpetuity,

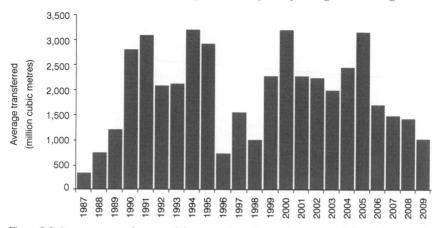

Figure 6.6 Average annual water rights transfer volumes in western USA for the period 1987–2009

Table 6.5 Water transfer prices in western USA 1987–2007 (adapted from Donohue, 2009)

Sectors	Sales (US$)		Leases (US$)	
	Mean price	Median price	Mean price	Median price
Agriculture-to-agriculture	2,362 (1.91)	1,451 (1.18)	36 (0.029)	11 (0.009)
Agriculture-to-environment	2,565 (2.08)	552 (0.45)	126 (0.10)	32 (0.026)
Agriculture-to-urban	4,552 (3.69)	2,896 (2.35)	424 (0.34)	56 (0.045)

Note: numbers not in brackets are prices per acre-foot; numbers in brackets are prices per m^3

command much higher prices than short-term leases. Also, prices are highest for transfers from agriculture to urban supply and lowest within the agricultural sector. The large differences in prices across different sectors may suggest, among other factors, that water markets have not yet fully developed (Libecab, 2010).

Until 1999, increases in median sales prices were modest, from under US$ 2,000 per acre-foot (US$ 1.62 per m^3) in 1987 to about US$ 4,000 per acre-foot (US$ 3.24 per m^3) in 1999. After 1999, water sales prices to urban supplies outstripped those to agriculture. In 2003, sale prices to urban supplies reached a peak of nearly US$ 14,000 per acre-foot (US$ 11.35 per m^3) while those to agriculture reached a low of US$ 1,000 per acre foot (US$ 0.81 per m^3). In 2007, the two sale prices were approximately the same, about US$ 6,500 per acre-foot (US$ 5.27 per m^3). It is worth noting that sales data have been largely based on transactions in Colorado where the institutional structure of the Colorado Big-Thompson scheme and the sole irrigation district within the water scheme make transfers much easier (Donohue, 2009). Changes in median lease prices from agriculture to agriculture remained more or less the same over the period, at about US$ 10 per acre-foot (US$ 0.008 per m^3). Lease prices for transfers from agriculture to urban supply were around US$ 50 per acre-foot (US$ 0.04 per m^3), but with fluctuations of up to US$ 400 per acre-foot (US$ 0.32 per m^3).

Water banks

The first water bank in the western USA was established in 1979 in Idaho, although rental pools have existed since 1932. There has been some groundwater trading activity through the Idaho Water Supply Bank. In fact, 20–50 per cent of transferred groundwater rights are facilitated by the bank (Contor, 2010). Since the mid-1990s water banks have been established in most states of the western USA, mainly concerned with surface water. Groundwater banking is relatively new. According to Clifford *et al.* (2004) its development has been hampered by the lack of a definite allocation system specifying the quantity available for transfer. Allocation proposals to assist groundwater banking put forward by the same authors, included:

1   *Yield-stock rights*: In this system, individual water users are allocated property rights for a share of the groundwater.
2   *Unitisation or pooling:* This involves the management of an aquifer by a single entity, as for oil or gas, which allocates groundwater to individual owners of rights (Jarvis, 2011).
3   *Proportional rights:* In this mechanism, the groundwater rights market is based on the annual safe yield of the aquifer.

There are a number of groundwater banking programmes in the western USA. Their trading activity has, however, been generally limited. They include the Truckee Meadows Groundwater Bank in Nevada set up in 2000, the Edwards Aquifer in Texas set up in 2001, the Deschutes Water Exchange Groundwater Mitigation Bank in Oregon, set up in 2003, and a number of banking programmes in California. The best-known of these is the Semitropic Groundwater Banking Program. The program was established in 1958 to supplement groundwater resources in Kern County in the San Joaquin Valley. Surface water for agriculture after 1973 from the State Water Project (SWP) via the California aqueduct relieved the stress on groundwater. The development of a water banking programme as a potential mechanism for the continued improvement of groundwater levels was identified in 1992. The Semitropic Groundwater Bank was established in 1994, although the scheme started operating in 1990. Surplus surface water is provided to Semitropic by seven banking partners (originally, there were five banking partners from the agricultural, urban and private sectors) for irrigation instead of pumping groundwater (in-lieu recharge), and an equivalent volume of groundwater (less 10 per cent for losses) is credited as stored water to the banking partner. To a lesser extent, Semitropic utilises artificial recharge by allowing surface water in basins or ponds to infiltrate to the water table. The storage–recovery cycle is approximately 12 years, nine years to fill the shares purchased by the banking partners and three years for pump back. The original allocation was for a storage capacity of 1 million acre-feet (1,233 Mm^3) with an annual pump-back capacity of 90,000 acre-feet (111 Mm^3). This was increased in 2002 to 1.65 million acre-feet (2,035 Mm^3) with an annual pump-back capacity of 200,000 acre-feet (247 Mm^3) in order to enable

Semitropic to sign up additional partners to join the scheme or for the existing partners to increase their share of recovered water (US Bureau of Reclamation (USBR), 2007). In the 21-year period from 1990 to 2010, the total volume provided by the banking partners for storage was approximately 1,356,400 acre-feet (1,673 Mm³). Of this 666,588 acre-feet (822 Mm³) has been recovered and 689,813 acre-feet (851 Mm³) is estimated to have remained in storage. (Semitropic Water Storage District; n.d.). Recovery was mainly between 2007 to 2009, amounting to 452,932 acre-feet (559 Mm³) or 68 per cent of total recovery. Both the stored capacity and pump-back quantities have been much less than anticipated.

Dellapenna (2005) in his discussion on water markets commented on the California water bank. This is a small bank which was set up in 1992 in response to the severe drought in the late 1980s and early 1990s to transfer water from agricultural use to urban use. He pointed out that the state was the only buyer and only seller of water rights, that prices were set administratively, and that the state ignored third-party effects. In a more recent publication, he summed up his view on the Californian water bank as follows:

> This was not a market in any meaningful sense of the term, but the implementation of government policy through economic incentives, with at least a veiled hint of coercive power.
>
> (Dellapenna, 2012)

The case of the Edwards Limestone aquifer, Texas

The case of the Edwards aquifer is a good example of the transfer of groundwater rights to achieve resources sustainability and reduce adverse impacts. Information on the Edwards aquifer may be found at its website, www.edwardsaquifer.net.

The Edwards aquifer is a large limestone aquifer that serves the agricultural, municipal, recreational and household needs of over 1.7 million users. Importantly it also feeds springs that sustain diverse ecosystems. In the 1950s, due to a severe drought, there were concerns about groundwater resources. As a result, the Edwards Underground Water District was formed, to protect and conserve the aquifer, but as it had no authority to restrict groundwater abstraction, it operated largely as a data collection agency. The Texas Water Plan in 1968 concerning the Edwards aquifer concluded that based on historical rates of recharge and discharge, abstraction from the aquifer should not exceed 400,000 acre-feet per year (493 Mm³ a⁻¹). Following a lawsuit in 1991 against the US Fish and Wildlife Service alleging that, contrary to federal legislation, the service was not adequately protecting endangered species that depended upon the aquifer, in 1993 the federal court issued a ruling requiring the state of Texas to implement a regulatory plan to limit abstractions from the aquifer to a sustainable level and maintain the ecologically important spring flows.

The Edwards Aquifer Act was enacted in May 1993 (with amendments in 2001 and 2005) and provided for the establishment of the Edwards Aquifer Authority (EAA) and the management of the Edwards aquifer. The EAA began operations in June 1996, as a 'conservation and reclamation district', to manage the southern portion of the Edwards Aquifer. The Act required that abstractions from the aquifer should not exceed 400,000 acre-feet per year ($493 \, \text{Mm}^3 \, \text{a}^{-1}$) after 1 January 2008 (in 2006 the cap was 450,000 acre-feet per year ($555 \, \text{Mm}^3 \, \text{a}^{-1}$) raised to 572,000 acre-feet per year ($706 \, \text{Mm}^3 \, \text{a}^{-1}$) in 2007), and defined trigger groundwater levels at an index well and springflows for the management of the aquifer and pools (EAA, 2010). The Act prohibited any person from withdrawing water from the aquifer except as authorised by a permit issued by the EAA or by the Act, each permit specifying the maximum rate and total volume that could be extracted in a calendar year. Until then the doctrine of Absolute Ownership applied, which allowed a land owner to abstract as much water as he liked (but see Chapter 3 with reference to the ruling of the Texas Supreme Court decision of 24 February 2012 in the *EAA* v. *Day and McDaniel* case that 'land ownership includes an interest in groundwater in place that cannot be taken for public use without adequate compensation'). Existing users were to be granted an initial regular permit by filing a declaration of historical use of groundwater during the period from 1 June 1972 to 31 May 1993. Additional permits would be added, subject to the cap of 400,000 acre-feet per year ($493 \, \text{Mm}^3 \, \text{a}^{-1}$). If after January 2008, the overall volume authorised to be abstracted was greater than the cap volume, the maximum authorised withdrawal was to be reduced by an equal percentage as is necessary to reduce overall abstraction to the cap volume. The EAA was also authorised to acquire permitted water rights for a number of purposes, including holding these rights for resale or retirement as a means of complying with pumping reduction requirements. The cost of reducing withdrawals or permit retirements to 450,000 acre-feet per year ($555 \, \text{Mm}^3 \, \text{a}^{-1}$) by 31 December 2007 was to be borne solely by users of the aquifer, and from this level to 400,000 acre-feet per year ($493 \, \text{Mm}^3 \, \text{a}^{-1}$) by 1 January 2008, equally by aquifer users and downstream water rights holders. With regard to transferability of water rights, the Act required: (a) that water withdrawn from the aquifer must be used within the boundaries of the EAA; (b) that the EAA by rule may establish a procedure by which a person who installs water conservation equipment may sell the water conserved, and (c) a permit holder may lease permitted water rights, but a holder of a permit for irrigation may not lease more than 50 per cent of the irrigation rights initially permitted. The user's remaining irrigation water rights must be used in accordance with the original permit and must pass with transfer of irrigated land. Presumably this last condition has to do with irrigation return flows to the aquifer, which form part of its recharge, and would be lost if all water was for water supply, unless of course the resulting sewage effluent was treated and returned as recharge.

Applications for permits exceeded by far the proposed cap, and this together with several legal challenges requiring the amendment of the EAA's rules, delayed things further, so that the first permanent permits were not issued until 2001. Initial regular permits (i.e. those based on use from 1972 to 1993) were granted to 822 out of 1,085 applicants (632 for irrigation, 161 municipal, 211 industrial, and 81 domestic/livestock) corresponding to a volume of 650,000 acre-feet per year (801 $Mm^3 a^{-1}$). In November 2005, the EAA issued its final order establishing abstraction permit amounts. Until then all permits were temporary and provisional. It recognised rights to pump 548,884 acre-feet per year (677 $Mm^3 a^{-1}$), which is 98,884 acre-feet per year (122 $Mm^3 a^{-1}$) in excess of the legislatively mandated cap of 450,000 acre-feet per year (555 $Mm^3 a^{-1}$). It was suggested that the excess should be dealt with by treating a percentage of each user's rights as 'junior' rights and interruptible if the aquifer level or spring flows were to fall below certain trigger levels. In 2006, the EAA declared mandatory groundwater restrictions to abstraction permit holders to allow groundwater to recover from a very severe drought. At the same time it introduced a fee rebate programme, estimated at US$ 2.2 million, which gave a financial incentive to non-agricultural abstraction permit holders to recoup their aquifer management fees for authorised groundwater volumes they did not abstract. In addition, it introduced a programme giving compensation to landowners for the removal of ash-juniper and brush control, in order to reduce evapotranspiration and increase recharge in the recharge zone of the aquifer (EAA, 2006). Other protection measures included various purchases of land between 2000 and 2007 by the San Antonio City Council mostly in the aquifer recharge area and reductions of building cover by 30 per cent, both measures for the purpose of conserving recharge to the aquifer. In the period 2000–2008 some 24,000 acres have been bought at a cost of US$ 22.8 million.

In the period 1997–2007, the total mean groundwater withdrawal from the Edwards aquifer was 404,600 acre-feet per year (499 $Mm^3 a^{-1}$) comprising: 91,200 acre-feet per year (112.5 $Mm^3 a^{-1}$) for irrigation, 280,300 acre-feet per year (345.7 $Mm^3 a^{-1}$) for municipal and domestic supply and 33,100 acre-feet per year (40.8 $Mm^3 a^{-1}$) commerce and industry (EAA, 2010). The mean flow from the springs was 569,700 acre-feet per year (702.7 $Mm^3 a^{-1}$). Interestingly, withdrawals in 2007 were the lowest at 296,900 acre-feet per year (366 $Mm^3 a^{-1}$) and spring discharge the highest at 620,600 acre-feet per year (765.5 $Mm^3 a^{-1}$).

The Edwards aquifer case illustrates that the legal process of making changes in water rights allocation can be lengthy and eventful. Limiting the total permitted abstraction to a sustainable level may take many years. Retirement of water rights or the giving of financial incentives to water rights holders to reduce abstraction can be costly. Aquifer management fees for municipal and industrial users in the affluent Edwards aquifer area have risen from US$ 25 per acre-foot (US$ 0.02 per m^3) in 2002 to an average of US$ 38 per acre-foot (US$ 0.03 per m^3) in the period 2005 to 2011. In 2012, there was a 20 per cent increase to US$ 47 per acre-foot (US$ 0.038 per m^3), mainly to cover for losses in 2010 as a result of rebates to non-agricultural users for conserved or unused water. Agricultural fees have been kept at US$ 2 per acre-foot (US$ 0.0016 per m^3) (EAA, 2012). The poor inhabitants and farmers in the rural agricultural sectors of developing countries may have difficulty funding groundwater protection measures through user fees.

Summary

To date the great majority of water rights transfers in western USA has been in surface water. They have been made possible by the development of large water transfer engineering schemes funded by government. Permanent bulk transfers of groundwater rights across user sectors are not known.

In the last 23 years (1987–2009) the reallocation of water resources by means of water rights trading or through water banks has been very small, less than 5 per cent of the total water used. Although there has been a general upward trend in water rights trading primarily in Colorado and Texas, the number of transactions has remained small, on average less than 200 per year. More than 50 per cent of all trades have taken place in Colorado as a result of the Colorado-Big Thompson scheme. Transfers occur mainly within agriculture or from agriculture to urban supply, and recently from agriculture to the environment. Despite the fact that water rights trading has been taking place for many years, there is no evidence of a market of many buyers and many sellers having developed. On the contrary, there has been a tendency for water rights to accumulate in the hands of a few organisations. Also, water prices have not shown convergence but rather significant variations exist from state to state. Water banks and water rights transfers have been largely administered by the state and one may be forced to incline towards the view of Dellapenna (2005) that 'it is hard to see the invisible hand of the market place in any of this'. Groundwater has generally remained overexploited and government has had to intervene to address the depletion of aquifers, falling water levels and adverse impacts on the environment.

Australia

Water resources

Australia is divided into six states and various territories including, Western Australia in the west, Queensland, New South Wales, Victoria, the Australian Capital Territory in the east, Northern Territory and South Australia in the centre, and the island state of Tasmania in the southeast. Most of Australia (more than 75 per cent) is arid (desert) or semi-arid. It is tropical and subtropical along the northern and northeastern coasts, and temperate along the southwestern and southeastern coast and Tasmania. The 'normal' long-term (1961–1990) average annual rainfall for the whole of the Australia is 472 mm. The central, southern and western parts have an average annual rainfall of less than 200 to 300 mm. Rainfall generally increases towards the coastal areas to 1000–1200 mm along the southwestern coast (Perth area), 1000–1600 mm on the northern coast, and 1000 mm to more than 2000 mm along the eastern coast and Tasmania. Along the western coast rainfall remains low, less than 300 mm (Bureau of Meteorology, n.d.). The period 2001–09 was marked by a drought. Annual rainfall was generally less than the long-term average (in 2002 and 2005 it was 339 mm and 399 mm, respectively), especially in the centre, west and southwest, and also in the Murray–Darling Basin in the southeast.

The average groundwater recharge in 2004–05 in the western, central and southern arid and semi-arid areas was 5 mm or less, increasing to about 10–13 mm in the northern and eastern coastal areas and southwest coast. In the Murray–Darling basin it was approximately 9 mm and in Tasmania 23 mm (NWC, 2007). Recharge in 2004–05 was generally less than the mean recharge, except in the southwest coastal area where it was a small percentage higher. In the central areas, it was 50–70 per cent of mean recharge, in the coastal areas, including Tasmania, 70–90 per cent, and in the Murray–Darling Basin it was 90 per cent.

The total inflow to Australia's water resources in 2004–05 was approximately 292,000 Mm^3 or 10 per cent of the rainfall of 364 mm (NWC, 2007). Of this approximately 242,800 Mm^3 or 83 per cent was surface water runoff and 49,200 Mm^3 or 17 per cent deep drainage into aquifers (groundwater recharge). Owing to the drought, total inflows in 2004–05 were approximately 20 per cent less than in 1996–97. The total sustainable yield for 2004–05 for the whole of Australia was given by NWC (2007) as 55,708 Mm^3 comprising 37,842 Mm^3, or 68 per cent surface water, and 17,866 Mm^3, or 32 per cent, groundwater. The total consumptive use was 34 per cent of the total sustainable yield.

In 2009–10, the estimated total inflow (landscape water yield) for the whole of Australia was 96 mm (equivalent to approximately 736,574 Mm^3) or approximately 14 per cent of rainfall for the year (Bureau of Meteorology, 2011). This is about two and half times greater than the 2004–05 average, and 40 per cent greater than the long-term average. The accessible capacity of surface water storage for June 2010 was 40,500 Mm^3. The Bureau of Meteorology did not provide any

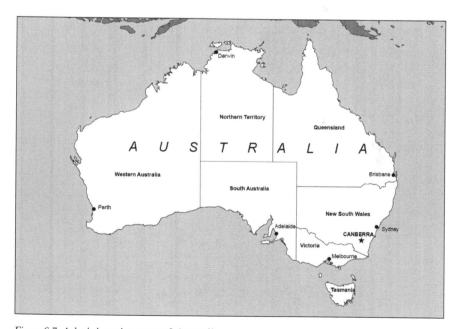

Figure 6.7 Administrative map of Australia

estimates of groundwater status 'due to the limited amount of quality-controlled data available in suitable form' but indicated that in some of the aquifers of the Murray–Darling basin and in the South Australian Gulf Region groundwater levels were falling (Bureau of Meteorology, 2011).

Groundwater constitutes a significant source of water for consumptive use in Australia. In Western Australia and the Northern Territories groundwater is the main source (NWC, 2011b). According to the National Land and Water Resources Audit (NLWRA, 2001), Australia has 25,780 Mm³ of groundwater that can be extracted sustainably each year for potable, stock and domestic use, and irrigated agriculture. This estimate was revised down to 17,866 Mm³ for 2004–05 (NWC, 2007). Table 6.6 shows the long-term average groundwater sustainable yield in each state and territory together with the mean long-term rainfall.

The sustainable groundwater yields and abstractions in 51 'priority geographic areas' (these were areas nominated by the states and territories where detailed water resources assessments were undertaken) for 2004–05 are shown in Table 6.7. The table suggests that in general groundwater abstraction is significantly lower than the sustainable yield in all areas. The exception is the Great Artesian Basin where it is about 16 per cent higher. The Great Artesian Basin stretches across three states (Queensland, South Australia and New South Wales). It is one of the world's largest aquifer systems and stores 8,700,000 Mm³ of groundwater. At its centre it is a more than 3,000 m thick. Uncontrolled artesian flows from old unsealed boreholes have contributed to groundwater losses. A programme to rehabilitate old boreholes (the Great Artesian Basin Sustainability Initiative) which started in 1999 has resulted in a reduction in wastage. During phase 2 of the programme water wastage was reduced by approximately 23 Mm³ a⁻¹ (Sinclair Knight Merz, 2008). In the Murray–Darling Basin groundwater levels have been falling in a number of areas and aquifers where development has been taking place suggesting that groundwater is being overabstracted (Murray–Darling Basin Commission, 2008).

The lower rainfall in the period 2001–2009 together with temperature increases resulted in a decrease in available resources, which was reflected in a reduction of water consumption by almost two-thirds, from 21,703 Mm³ in 2000–01 to 18,767 Mm³ in 2004–05 and 14,101 Mm³ in 2008–09 (ABS, 2010). The main reduction of more than 50 per cent (from 14,989 Mm³ in 2000–01 to 6,996 Mm³ in 2008–09) was in agriculture (mainly irrigation). In 2000–01 agriculture represented 69 per cent of total water use; by 2008–09 it had reduced to just below 50 per cent. Reductions in agriculture have been mainly in the Murray–Darling Basin, which accounts for approximately 50–60 per cent of total water use. In the same period, consumption in other sectors generally increased, in urban water supply from 2,420 Mm³ to 2,840 Mm³ and in industry, manufacturing and mining from 1,976 Mm³ to 2,512 Mm³. Household use in the rural areas, however, decreased from 2,278 Mm³ to 1,768 Mm³. Interestingly, although agricultural use had decreased expenditure increased by nearly 40 per cent (ABS, 2010).

Table 6.6 Long-term average groundwater sustainable yield 2004–05 (based on data from NWC, 2007)

State or territory	Area (km²)	Mean rainfall 1961–1990 (mm)	Groundwater sustainable yield (Mm³) (round figures)	Groundwater sustainable yield / area (m³ / km²)
Western Australia	2,522,482	566	3,678	1,458
Northern Territories	1,344,875	548	2,389	1,776
South Australia	979,330	236	1,253	1,279
Queensland	1,725,594	630	322	187
New South Wales	804,059	566	7,164	8,910
Australian Capital Territory	2,363	566	7	2,962
Victoria	226,846	654	521	2,297
Tasmania	67,096	1,168	2,531	37,722
Total/average	7,672,645	472	17,866[a]	2,328

a   As given by NWC, 2007

Table 6.7 Sustainable groundwater yield and abstraction in 51 'priority geographic areas' for 2004–05 (Source: NWC, 2007)

Geographic area/basin	Area (km2)	Groundwater sustainable yield (Mm³) (round figures)	Groundwater abstraction (Mm³) (round figures)	Groundwater abstraction as percentage of sustainable yield (per cent)
Western Australia	2,522,482	663	367	55
Northern Territories	1,344,875	570	34	6
South Australia	979,330	762	287	38
Queensland	1,725,594	117[a]	270[a]	–
New South Wales	804,059	1,018	840	83
Australian Capital Territory	2,363	7.455	0	–
Victoria	226,846	107	82	77
Tasmania	67,096	172	3	2
Great Artesian Basin	1,711,000	475	549	116
Murray–Darling Basin	1,061,469	1,625	1,490	92

a   The sustainable yield does not include the Condamine–Ballone area which is part of the Murray–Darling Basin. Abstraction in this area is approximately 141 Mm³.

Water legislation

Before colonisation, water rights were part of communal property rights held by the indigenous aboriginal populations. European colonisation began with the establishment in 1788 of the British penal colony of New South Wales. Two more British colonies followed: Western Australia in 1829 and South Australia in 1836 (it became a crown colony in 1842). Victoria, Queensland and the Northern Territory were initially part of New South Wales. In 1901, Australia became a federation of the six colonies. The British colonists brought with them their own systems of riparian rights for surface water and absolute ownership for groundwater. However, as in the dry western USA, the riparian system was found to impede economic development (Musgrave, 2011), and both this and the absolute ownership doctrine were replaced by a licensing and allocation system (McKay, 2011). The 1880 Water and Conservation Act reduced the riparian rights of individuals in the interests of the state, while the 1886 Irrigation Act (Victoria) tied water allocations explicitly to land and vested the ownership of water in the British Crown (Godden, 2005; Musgrave, 2011). The other states soon followed Victoria's lead. In practice, groundwater continued to be extracted without any control, allocations were at will and incentives to individuals to moderate their use were weak (McKay, 2011). As a result, individual state governments introduced several cross-border agreements, such as the Great Artesian Cooperative Agreement to cap overflowing boreholes and the Murray–Darling Basin agreements. These agreements were the forerunner of later federal participation in the management of the water resources of the commonwealth. In 1994, the Council of Australian Governments (COAG, 1994) introduced reforms to overcome the degradation of surface water and groundwater resources. These included the separation of water rights from land rights so that water rights could be traded, and the allocation by states of water for environmental purposes. Fundamental to the reforms was the requirement for environmentally sustainable development (ESD) and also more private sector participation in water management.

The 1994 reforms produced a flurry of water rights applications from landowners who had water licences attached to their land, which either they never used ('sleepers') or had ceased to use them ('dozers'). This resulted in water claims exceeding sustainable yield, which negated somewhat the intention of the reform to reduce abstraction.

The complexity of the 1994 reforms led to a second series of major reforms initially announced at the 2003 COAG meeting (COAG, 2003), and later in a statement in 2004 (COAG, 2004) referred to as the National Water Initiative (NWI). The NWI was a national effort aimed at achieving a number of objectives, among them, an increase of the country's efficiency in water use, greater certainty for investment and productivity, and expanding the trade in water in the different states. ESD principles also formed part of the NWI in so far as it recognises the importance of ensuring the health of river and groundwater systems, including the establishment of clear pathways to return all systems to environmentally sustainable levels of extraction. The National Water Commission Act 2004 (expiring on 30 June 2012) established the National Water Commission (NWC) to assist with the implementation of the NWI. The intergovernmental agreement on NWI was considered by COAG in 2004 and was signed by all states by 2006.

The 2007 Water Act

An important piece of recent legislation is the 2007 Water Act ('An Act to make provision for the management of the water resources of the Murray–Darling Basin, and to make provision for other matters of national interest in relation to water information, and for related purposes'), which, as its full title implies applies not only to the Murray–Darling Basin, but also in other areas nationally. The Act requires that, *inter alia*:

1   water resources in the basin are managed in the national interest;
2   water resources that are overallocated or overused are returned to sustainable levels of extraction;
3   the ecological values and ecosystems of the basin are protected and restored, this particularly in relation to the impact of abstraction; and
4   economic returns from the use of and management of the water resources of the basin are maximised.

Water rights may be traded between states and, pursuant to this trade, water can be delivered between states. According to the Act, the trading and transfer of water rights will promote the more efficient use and continued availability of water resources, the health of the environmental assets associated with the basin water resources and the economic and social wellbeing of the communities of the basin. The Water Act provided for the establishment of the Murray–Darling Basin Authority (MDBA) charged with the development of the basin plan for the integrated management of water resources in the basin and the assignment of responsibilities, including the provision of advice on water sharing and water market and water trading rules to the Australian Competition and Consumer Commission (ACCC). In 2008 there was an Intergovernmental Agreement on the Murray Basin Reform with state governments referring state power to the commonwealth government. During 2008–2011, a number of initiatives took place aimed at assisting further the reforms on water rights trading.

Recent state water legislation

In the last 15 years, most states have enacted new or revised old legislation to deal with the management, allocation and protection of water resources, including regulations to safeguard the environment. There are no specific provisions in legislations for social issues. However, customary and domestic rights, and the rights of indigenous people are protected by legislation. Historically, European settlers appropriated water resources without consultation with the indigenous aboriginal populations. Stakeholder participation in recent legislation includes arrangements for consultations with communities, such as the establishment of advisory committees, the direct representation of communities on allocation decision-making bodies, public invitations to make submissions in the development of water resources plans, and public meetings. Transparency is encouraged and in recent years there has been a plethora of freely available publications, many via the internet, of water resources availability and use, environmental impacts, water resources plans, water rights trading, etc.

Water rights in different states come under different names, including: entitlements, water entitlements, water access entitlements, share component, water allocation, water licences, take-and-use licence, extraction licence. The two main categories are water access entitlements and water allocations. Water access entitlements refer to a perpetual or ongoing entitlement to exclusive access to a share of water from a specified consumptive pool (NWC, 2011a). Water access entitlements may vary in terms of security. A water allocation is a specific volume of water allocated to water access entitlements in a given season.

Table 6.8 presents some of the main relevant characteristics of water rights in different states. In the past, the quantity of water for agricultural use permitted by the water right was based on the area to be irrigated, which led to excessive use. Gradually (from the late 1960s to the early 1980s) state legislation replaced this with volume-based licences. The definition and measurement of volume vary from state to state. Characteristically water access entitlements are related to a share of the available resource. The duration of the right varies but it is generally 10–15 years or in perpetuity. In most states, both the volume and duration of the water right may change subject to hydrological conditions, impacts on the environment or interstate agreements. Stock and domestic rights are generally transferable with land and are perpetual. Where water rights are transferable it is usually subject to the approval of the minister or other responsible body. The link between water rights and land rights still remains in some states. It is the objective of the government to have a national water system in which water rights can be traded independently of land.

Water rights transfers

Since the mid-1990s, the Australian government has expended considerable financial resources and effort on reforming the water sector and preparing it for an economic approach to the management and allocation of water resources. The benefits have been substantial, particularly in the better understanding of the nature of water resources, their utilisation and availability, and their contribution to the environment. The relatively speedy introduction of new legislation and the concerted efforts in the assessment of water resources have been especially impressive. The unbundling of water rights and the introduction of formal, explicit and transferable water rights should prove of value to the future management of water resources, irrespective of any potential future benefit arising from economic efficiency. To a certain extent, transferable water rights have proved useful, although at a cost, in enabling government to buy back irrigation water rights in order to ensure sustainability and protect the environment. However, it may be argued that the proliferation of water rights has been the result of the COAG 1994 reforms.

Table 6.8 Some of the characteristics of water rights in Australia

Geographic area	Volume and measurement	Duration	Transferability and water trading	Legislation
Western Australia	Licence to take water. Water entitlement is defined by annual licence volume. Use of meter to measure flow.	For a fixed period of time or for an indefinite period.	The holder of a licence may transfer the licence to another person, but the person must be an owner or occupier of the land to which the licence relates. Or he may transfer his entitlement to a person who is eligible to hold a licence of the same kind. Trading: for intrastate trading buyers and sellers to submit application in paper form to the Department of Water for approval. Interstate trading is not permitted.	Rights in Water and Irrigation Regulations 2000. Rights in Water and Irrigation Act 1914. Water Resources Management Bill being drafted to consolidate existing acts into one legislation.
Northern Territories	Extraction licence. Annual licensed volume. As approved by the Controller.	Not more than 10 years (groundwater). For longer if approved by minister for special circumstances.	Water right transferred with land. For intrastate trading buyer and seller to submit paper form to the Department of Natural Resources, Environment the Arts and Sport (NRETAS). Trade application made available to public for comment before approval by NRETAS. To date there have been no water trades.	Water Act 2009.
South Australia	Take licence. Water access entitlement: an entitlement of share of water available in the consumptive pool to which the licence relates; subject to reduction by the minister in relation to water resource and water quality changes or to prevent damage to ecosystem in relation to the Murray–Darling Basin agreement, if relevant.	Not stated (perpetual)	A water licence is personal property and may be passed to another person. Also, the holder of the licence may transfer a water access entitlement or part of it to another person. Transfer requires the approval of the minister. Water trading is allowed. Trade applications lodged with the Department of Water for approval. For interstate trade applications must be lodged with both the South Australia and interstate approving authorities. Water rights are expected to be fully unbundled by 2013.	Natural Resources Management Act 2004.

Continued...

Table 6.8 continued

Geographic area	Volume and measurement	Duration	Transferability and water trading	Legislation
Queensland	Resource or distribution operations licences are granted to meet future water requirements. The holder of a licence to install meter and carry out reporting and monitoring. The licence can be amended following amendment of resource operations plan. Volumetric limit is the maximum volume of water in megalitres that may be taken under the allocation	For licences not subject to a Resource Operations Plan for up to 10 years. For licences subject to a Resource Operations Plan for a specified period. For water allocations subject to a Resource Operations Plan ongoing, Resource Operation Plans are subject to 10-yearly reviews.	Water allocations that are subject to Resource Operations Plan are transferable independent of land. Water licences are not transferable independent of land , irrespective of whether or not they are subject to a Resource Operations Plan. Water trading is administered by the Queensland Department of Environment and Resource Management (DERM). Intrastate leases or transfers do not require approval from DERM, but some water trades need approval. Intrastate trade between Queensland and New South Wales remains limited to specific circumstances until the New South Wales sharing programme has been finalised.	Water Act 2000
New South Wales	Water access licence (includes aquifer access licence and utility licence) entitles holder to specified shares in available water or source expressed as maximum volume, proportion of available water or as specified in units (volume per year for water utilities). Minister can amend the share or extraction components of the access licence. Water access licence also required for mining activities	Major or local water utility 20 years. Local water utility licence to be varied at the end of 5 years. Other licences not greater than 10 years.	Transfer of water access licence is allowed between utilities. Transfer between persons is allowed for periods of not less than 6 months (term transfers). Interstates transfers require minister's approval. Organisations dealing with water trades and transfers are: Office of Water, State Water and Land and Water Management Authority Intrastate trades and transfers must be registered on the Water Access Licence Register. Interstate trades must be notified to the Murray–Darling Basin Authority.	Water Management Amendment Act 2010; Water Management Amendment Act 2008; Water Management Act 2000; Water Act 1912

Geographic area	Volume and measurement	Duration	Transferability and water trading	Legislation
Australian Capital Territory	Licence to take water. Water access entitlement stated as the lesser of a percentage of the total amount of water available for taking from time to time in the water management area or a stated maximum volume. Water licences issued if person holds a water access entitlement.	Water licence issued for the term stated in the licence. Allocations subject to a ten-yearly review.	Water access entitlements may be transferred on approval by the Water Authority for units greater than 0.5 megalitres. Licences to take water licences are not transferable. Environment Protection Authority (EPA) is responsible for licence transactions. Water trading has only occurred through transfer of land.	Water Resources Act 2007.
Victoria	Take-and-use licence: volume defined as the maximum amount that may be taken in particular periods or circumstances or payment for the amount of water used; installation and use of devices to measure flow. Environmental allocations: the means for quantifying amount to be defined. Bulk allocations to authorities to be granted by minister. Licence to take water must include an annual use limit licence volume.	Duration of licence no more than 15 years. For groundwater where the yield of an aquifer is uncertain or there may be a potential impact due to groundwater extraction no greater than three years.	Transfer of water licence only to a person who owns or occupies the land specified in the licence. Environmental allocations and bulk entitlements transferable after approval by minister. Intrastate transfer of water access entitlement has to be approved by the water corporation before the buyer and seller can settle the transaction. For interstate transactions, approval is required from the water authority in both the state of origin and the state of destination.	Water Act 1989; Policies for Managing Take and Use Licences; Water (Resources Management) Act 2005.

Continued...

Table 6.8 continued

Geographic area	Volume and measurement	Duration	Transferability and water trading	Legislation
Tasmania	Water allocation of a licence may be granted in accordance with any relevant water management plan or in accordance with the objectives of the Water Management Act. It may be reduced to give effect to water management plan. May be fixed by specifying the volume of water that may be taken and used.	Licence remains in force for as long as the minister determines and as specified in the licence.	A water licence including the right to water allocation endorsed on it is the personal property of the licensee and is alienable in accordance to the provisions in the Water Act and any other law relating to the passing of property. Licences and water allocation are transferable, subject to the approval of the minister. Irrigation rights may be transferred after approval of the undertaker. The selling of water or the transfer of an entitlement involves the submission of a hard copy form to the Department of Primary Industries, Parks, Water and the Environment (DPIPWE) for approval.	Water Management Act 1999; Irrigation Clauses Act 1973.

Note
Information based on the state Water Acts; also information from http://www.nationalwatermarket.gov.au

In 2007, the Australian government allocated AUS$ 3.1 billion (approximately US$ 2.6 billion) to buy back 1,500 Mm³ of existing Murray–Darling Basin (MDB) water entitlements from willing sellers. In the southern MDB, government buybacks from irrigators, mostly in 2009–10, amounted to about AUS$ 1 billion (approximately US$ 0.79 billion) in return for 672 Mm³ entitlements (NWC, 2012).

Uncertainty and risk

The first ten years of the new millennium were marked by a serious and persistent drought, which brought to the fore nature's quixotic behaviour in dispensing its largesse, and with it the difficulties of determining available resources with a certainty that water trading can depend upon. Risk and uncertainty in water management in Australia were discussed by Quiggin (2011), mainly in relation to the Murray–Darling Basin. Variations in annual and seasonal rainfall, climate change, demand and government policy were given as sources of uncertainty. Groundwater requires much study and investigation, and often long-term monitoring, in order to reach a reasonable level of certainty of its resource potential. Even so, temporal variations remain difficult to predict. Droughts, which have afflicted all continents at one time or another, are still not possible to predict with any degree of certainty, despite all the recent advances in meteorology. The attenuation of water rights under the NWI, which is reflected in the allocation of an annually determined share of the consumptive pool rather than a right to a fixed volume of water in perpetuity (McKay, 2011), is probably in response to this uncertainty. The 2004 COAG meeting assigned the risk of future reductions in water allocations as follows (COAG, 2004; Quiggin, 2011):

1   reductions arising from natural causes, such as climate change, drought or bushfire, to be borne by water users;
2   reductions arising from bona fide improvements in knowledge about water resources availability to be borne by water users up to 2014; after 2014, water users to bear this risk for the first 3 per cent reduction in water allocation and the relevant state or territory and the Australian government above 3 per cent;
3   reductions arising from changes in government policy not previously provided for to be borne by the Australian government;
4   where there is voluntary agreement between a state or territory and key stakeholders, a different risk model may be implemented.

Cap and trade

The realisation that water resources were scarce led to the concept of 'cap and trade'. The cap represents the limit of the resource available for extraction (consumptive pool), consistent with sustainable levels of extraction (NWC, 2011a). Individuals are provided with entitlements to a share of the total pool. Trade refers to the buying and selling of water access entitlements and water (seasonal) allocations. Thus, in theory, water is reallocated to its more economically efficient use, while ensuring environmental sustainability. As already indicated, environmentally sustainable development was fundamental to the NWI reform. As a result, the environment and long-term resource sustainability were given prominence in the allocation of the resource pool. Nevertheless, during the drought, the market proved inadequate to safeguard the environment, and government had to resort to buybacks of water access entitlements in an attempt to reduce exploitation to sustainable levels. In 2007, the Australian government allocated AUS\$ 3.2 billion (approximately US\$ 2.6 billion) to buy back 1,500 Mm³ of existing MDB water entitlements from willing sellers. In the southern MDB, government buybacks from irrigators, mainly in 2009–10, amounted to about AUS\$ 1 billion (approximately US\$ 0.79 billion) in return for 672 Mm³ of entitlements (NWC, 2012).

Water marketing in Australia continues to be intensely promoted as a means of achieving economic efficiency, the sustainable use of water resources and protection of the environment. Nevertheless, it has been felt necessary to maintain, and in some cases add to, existing regulatory measures. This is somewhat contradictory to the free market model of no government intervention, but it is not at variance with experiences in other countries, including Chile, where water rights are freely tradable.

Groundwater rights transfers

Groundwater entitlements in Australia amount to 49 per cent by number and 21 per cent by volume of all entitlements (NWC, 2011b). The largest volume of groundwater entitlements is in New South Wales, closely followed by the arid Western Australia. The transfer and trading of groundwater rights has been limited. A snapshot of the level of groundwater trading is presented in Table 6.9. In the period 2010–2011, the average number of traded entitlements and allocations in all states was 0.6 per cent and 0.3 per cent, respectively, of all entitlements. In terms of volume, water entitlements trades were 2.3 per cent of all entitlements and allocation trades 0.8 per cent. There were no trades in the Northern Territories and Tasmania, no entitlement trades in Queensland and no allocation trades in the Australian Capital Territory. Allocated trade volumes are significantly less than the volume of entitlement trades, reflecting the limitation of the availability of resources. The largest traded volumes were in New South Wales, representing approximately 51 per cent and 64 per cent respectively of the total entitlement and allocation trades. Lower levels of groundwater trading of approximately 308 Mm³ for both entitlements and allocations were also identified for 2008–09, when again there was no trading activity in the Northern Territories and Tasmania (NWC, 2011b). Earlier reports also suggest small volumes of groundwater trade (Productivity Commission, 2006). The largest traded seasonal allocation volumes were reported in Queensland, 0.028 to 0.131 Mm³, and the largest entitlement volumes in South Australia, 0.029–0.089 Mm³. These traded volumes in relation to the total extraction rates are insignificant, in the range of 0.02 per cent to less than 0.001 per cent.

Various reasons have been suggested (Productivity Commission, 2006) for the poor trade in groundwater rights, among them:

Table 6.9 Groundwater entitlements and volumes traded in different geographic areas of Australia for 2010–11 (based on data from NWC, 2011b)

Geographic area	Groundwater entitlements (30 June 2011)		Groundwater trades in 2010–11			
			Entitlement trades		Allocation trades	
	Number (A)	Volume (B) (Mm³)	Number / per cent of A (per cent)	Volume (Mm³) / per cent of B (per cent)	Number / per cent of A (per cent)	Volume (Mm³) / per cent of B (per cent)
Western Australia	10,562	1,772	78/0.74	21.990/1.2	12/0.1[a]	8.066/0.5[a]
Northern Territories	255	131	0	0	0	0
South Australia	4,863	618	172/3.5	21.416/3.5	35/0.7	3.125/0.5
Queensland	15,264	1,038	0	0	35/0.2	2.328/0.2
New South Wales	74,197	1,934	151/0.2	74.362/3.9	155/0.2	31.192/1.6
Australian Capital Territory	144	1	3/2.1	0.004/0.4	0	0
Victoria	10,706	863	265/2.5	27.309/3.2	70/0.6	4.179/0.5
Tasmania	0	0	0	0	0	0
Totals	115,991	6,357	669/0.6	146.081/2.3	307/0.3	48.89/0.8

a Western Australia allocation trades relate to leases

1   hydrologically connected groundwater systems cover smaller areas;
2   inadequate knowledge of groundwater connectivity and levels of sustainable use in many regions;
3   entitlements in some regions not clearly defined and often significant regional differences in groundwater management;
4   entitlements to groundwater still linked to land in many regions.

Similar reasons have been given by the National Water Commission (NWC, 2011b). Again, the main reason was the limited hydrogeological connections of aquifers and limited physical infrastructure linking groundwater areas that lack hydrogeological connectivity. Administrative and management reasons have also been cited as reasons restricting groundwater trading activity, including: not fully unbundled groundwater rights, insufficient time for the market to develop, uncertainty with regard to the definition of groundwater management units (meaning the boundaries and extent of aquifers), groundwater rights still available on application in some areas and limited demand for groundwater where surface water is plentiful. Environmental constraints with regard to fears of further degradation of groundwater-fed wetlands and soil salinisation have also been significant factors in constraining groundwater transfers. Whilst all these factors have undoubtedly affected groundwater trading (although, hydrogeological connectivity does not seem to have affected the exploitation of groundwater in the Great Artesian Basin), it remains uncertain whether groundwater markets are likely to evolve substantially in the future, even if administrative constraints were to be removed. The examples from Chile, Mexico and the western USA also suggest a low level of groundwater trading. Even for surface water, which is in many respects an easier resource to evaluate and manage, trading has not been particularly significant as a proportion of total utilisation.

Surface water rights trading in the southern Murray–Darling Basin

The Murray–Darling Basin (MDB) covers much of southeastern Australia, extending across four states, namely Queensland, New South Wales, Victoria and South Australia. The MDB is largely agricultural. In 2005–06, it accounted for 65 per cent of Australia's irrigated land and 66 per cent of agricultural water use (ABS, 2008). All year round irrigation accounts for less than 15 per cent. Pasture for dairy farming is the main irrigated crop, followed by cereals (mainly rice), cotton, grapes, fruit and vegetables. Irrigated activities, particularly for horticulture, are better developed in the southern MDB, partly due to its physical conditions and partly to the regulated rivers that provide reliable access to water (NWC, 2011a). The surface water resources of the southern MDB are particularly well connected by means of reservoirs, canals and tunnels constructed mainly during the last century.

The southern MDB has been promoted as the primary example of the success of water rights trading in Australia. Over 90 per cent of Australia's water trading activity is concentrated in the southern MDB (NWC, 2011a). The growth of the water market, mainly within the agricultural sector, in the southern MDB has been attributed, at least partly, to the unique combination of physical characteristics (climate and soils), but especially to its large well-connected water systems. Yet, despite these natural advantages trading required concerted and ongoing effort for over 30 years (NWC, 2011a).

Water rights trading in the southern MDB started in the mid-1980s. Until 1994, the traded volumes were modest, generally less than 100 Mm^3 in seasonal allocations. The first significant increase occurred in 1994–95 to just under 800 Mm^3, the result of a large drop in water seasonal availability (NWC, 2011a). In 1995–96, seasonal allocation trading fell to approximately 500 Mm^3, and after this, it continued to generally increase reaching about 1,650 Mm^3 in 2009–10 and a maximum of about 2,700 Mm^3 in 2010–11 (NWC, 2011b). The drought has been a significant factor affecting water availability and prices, and therefore allocation trade volumes. Between 2001 and 2008, allocation trades remained low, on average approximately 910 Mm^3, with the lowest at 716 Mm^3 in 2006–07. The substantial increases in the wet years of 2009–10 and 2010–11 are a little puzzling but probably relate to increased allocations (more than 90 per cent after November 2011) and associated lower prices. After 2009 prices have been steadily decreasing from an average of approximately AUS\$ 340 per Ml (AUS\$ 0.34 per m^3 or US\$ 0.29 per m^3) in September 2009 to approximately AUS\$ 10 per Ml (AUS\$ 0.01 per m^3 or US\$ 0.01 per m^3) by May–June 2011 (NWC, 2011b). Although traded allocation volumes increased substantially in 2010–11, the number of trades decreased to a monthly average of approximately 670, in contrast to the previous three years of approximately 1,750 per month. This may reflect the opportunity taken by a small number of large buyers to purchase water at a low price. The proportion of allocations traded to announced allocation volumes has generally increased from 8 per cent in 2001–02 to 15 per cent in 2006–07. In 2007–08, it doubled to 30 per cent, but has remained at around this level since then.

Entitlement permanent trades in southern MDB have remained small, generally less than 100 Mm^3 until 2006. They generally increased after 2007 reaching a peak of approximately 1,355 Mm^3 in 2008–09, but fell again to approximately 550 Mm^3 in 2010–11. The increase in entitlement trades in 2008–2011 has been partly due to governmental environmental purchases (buybacks), 672 Mm^3 in 2009–10. The number of trades increased from 214 per month in 2007–08 to 448 per month in 2009–10, but decreased to 323 per month in 2010. Average prices for higher entitlement trades in the period 2007–2011 were approximately AUS\$ 1500–2000 per Ml (AUS\$ 1.5–2 per m^3 or US\$ 1.33–1.77 per m^3). Trading of water between states (interstate) has been generally small.

The proportion of surface water trades in relation to total entitlement volumes and diverted water remains small. For example the total volume of surface water entitlements in 2004–05 was approximately 14,600 Mm^3 and of surface water diversions 8,800 Mm^3 (NWC, 2007). In the same period the volume of allocation trades was approximately 831 Mm^3 and of entitlement trades 76 Mm^3. In percentage terms, allocation trades represented 5.7 per cent of the total volume of entitlements and 9.4 per cent of diversions, and entitlement trades 0.5 per cent of the total volume of entitlements and 0.9 per cent of diversions.

Environmental and social impacts of water rights trading

The environmental and social impacts of water rights trading are discussed mainly in relation to surface water in the MDB. As far as is known, groundwater trading has so far been too small to have had a significant impact on the environment or society. According to NWC (2012), groundwater trade volumes in the MDB to date have been small in comparison with extractions. However, NWC note that there may be a risk to groundwater ecosystems, even at the current low levels of trading.

The environmental, social and community concerns associated with water rights trading have been discussed by Bjornlund *et al.* (2012). With regard to the environment, the concerns relate to the potential adverse effects of the transferred water on water tables, river flows and river-water quality. Ecosystems may also potentially suffer from a decrease in their water supply due to the reactivation of previously unused water entitlements. State governments have introduced various restrictions to prevent adverse environmental impacts. These, however, have resulted in an increase in transaction costs reducing trading, especially in the entitlement market. In order to address this, Southern Australia in the 1997 Water Resources Act and later other states, unbundled the right to use the water from the ownership of the water entitlement.

The social concerns arising from water rights trading relate to the welfare of individual farmers, especially those not selling their water, and on the welfare of the wider rural community (Bjornlund *et al.*, 2012). Regarding the former the main issues are:

1   increasing maintenance and supply costs of water systems;
2   the closure of water supply channels, due to trading of water out of a supply system;
3   water rights hoarding and speculation resulting in price increases that farmers cannot afford;
4   invasion of weeds and pests into neighbouring properties from abandoned farmland.

In Victoria, the state government was forced to close water channels because trading moved much of the water out of some channels, leaving irrigators without water. The state government guaranteed compensation to irrigators for the loss of their property value.

With regard to the effect of water rights trading on the welfare of the wider community the main issues are:

1   the depopulation of the areas where irrigation farming has been abandoned;
2   a decline in social services (hospitals, schools, libraries, road maintenance) in rural areas;
3   a change in community composition.

At present, the federal and state governments may consider structural assistance to affected communities if the impacts become very serious. New structural adjustment policies are being planned for release in 2012, in conjunction with the proposed Basin Plan, by the Murray–Darling Basin Authority (MDBA, 2011).

The NWC discussed the impacts of surface water rights trading in the southern MDB in two reports (NWC, 2010; NWC, 2012). NWC considered that the buybacks of water entitlements from irrigators contributed beneficially to the environment by delivering water to wetlands and sites of environmental importance (environmental watering). However, rigorous evaluations of the outcomes of environmental watering are not yet possible, as the watering programmes have been only recently (mainly after 2008) established. Because many buyers are downstream of sellers (the major horticultural areas are downstream of rice and dairy areas), water has moved downstream, resulting in beneficial increases in summer flows during the drought. In general, eco-hydrological indices have suggested an improvement in river ecology, but these have been based on comparisons during the drought years when natural flows were low. Transmission flow losses along river courses due to increases in the distance that water has to travel (trading generally moves water downstream) in the period 1998–99 to 2008–09 are likely to have been small (NWC, 2010). However, with increasing trade downstream they have probably increased in 2009–10 (NWC, 2012). Increases in transmission water losses may have implications on resource availability and allocations. Modelling of trade-related changes suggested that the environmental effects of water trading are small in comparison with the impacts of drought and river regulation (NWC, 2010).

Social impacts have been explored by NWC (2010, 2012) in relation to irrigators who sold their entitlements to the commonwealth in buybacks. Irrigators who sold their entitlements have not necessarily been leaving the industry or their communities. Many irrigators have used the money to pay off or reduce debts while others have become more reliant on allocation trading and carryover trading, or have since repurchased entitlements. Water trading was not considered to have been responsible for the decreases in irrigated agricultural production in the dairy dominated regions of northern Victoria. NWC (2012), however, recognised that a broad range of irrigators and those directly employed in irrigation industries are apprehensive about the socioeconomic impacts of further buybacks.

Bjornlund *et al.* (2012) considered that at present, environmental impacts due to trading are difficult to prove. Entitlement trades have so far been too small to cause a real change whilst farm plans implemented by entitlement buyers have not been in place long enough to prove their worth. The moving of water from poor to better soils may reduce poor quality runoff to rivers. A study by URS (2010) reported that irrigation return flows had decreased significantly since 1993–94, with the most significant reductions occurring since 2000–01. They attributed this mainly to the drought, but also noted that improvements in irrigation efficiency and irrigation prices and in drainage collection were contributory factors. URS also indicated that trade was expected to affect return flows in the irrigation areas. Trade out would reduce return flows in the area and trade in increase them.

Similarly, Crase *et al.*, (2009) argued that more efficient irrigation practices and the trading of surplus may reduce return flows to rivers and the environment.

According to Bjornlund *et al.* (2012), population census data indicated that rural communities had shown resilience and maintained population and economic activity despite the export of water. Research by these authors suggested that irrigators have used the allocation market to manage scarcity and structural change. In an earlier publication, Bjornlund (2004) contended that markets promote sustainable farming practices, increase productivity, develop new employment options and allow farmers to remain in local districts after selling unviable properties. Frontier Economics (2007) indicated that water markets, in addition to facilitating structural change, provide land owners with more flexibility in production decisions enabling them to manage risks associated with water supply. There have been reports, however, of adverse social impacts, a disruption in local community life due to the influx of newcomers, loss of farming jobs and a significant dependence on off-farm work. Many farms, particularly dairy farms, with water no longer attached to them, remained abandoned (Fenton, 2006; Edwards *et al.*, 2008). Local inhabitants are unhappy with these changes and against water entitlement trading. Frontier Economics (2007) in a case study on the economic and social impacts of water trading, found that irrigation communities do not view water markets on their own, as fair or acceptable processes for allocating water. Mitchell *et al.* (2011) in their literature review of social research to improve groundwater governance indicated that concerns have been continually raised about the social impacts of water markets and trading in relation to the equity of the markets and the impacts on the wider community.

At present, there seems to be no firm conclusion on the impacts of water trading on rural communities and the environment. Even in the southern MDB where surface water trading has been prominent, it may take many years before any effects clearly emerge. Regarding groundwater trading, so far volumes appear to have been too small to have caused any identifiable effects.

Summary

The Australian government has done much for many years to encourage the development of water rights trading, including: changes in legislation to free water rights from land, studies to assess water resources, administrative measures to limit exploitation and direct action by buying water rights to reduce environmental impacts. Despite these measures, the market in groundwater rights remains feeble, 2 per cent or less of the total extracted groundwater, and in surface water in most states generally weak. Even in the southern Murray–Darling Basin which has had the great advantage of a very well-connected surface water system, the number of transactions remains small and traded volumes modest. In the last three years (2007–08 to 2010–11) water trades accounted for about 30 per cent of announced allocations, an increase of 15–20 per cent from previous years. However, when compared with the total volume of entitlements and diversions, traded volumes are probably small, less than 10 per cent for allocation trades and 1 per cent for entitlement trades. Publications by the National Water Commission are optimistic with regard to the future of water rights trading as a means of achieving sustainable use of water resources and economic efficiency. The experience so far is that physical factors (droughts and flooding) and external factors, such as the freeing of trade which allowed the import of cheaper agricultural products, seem to have had greater influence. Regarding groundwater, it is not possible to say at this stage whether or not trading will have a significant management role to play in the future.

Although much has been written about the economic achievements of water rights trading, the environmental and social impacts remain difficult to assess. This is partly because they generally take much longer to emerge and partly because, being less tangible, they are more difficult to evaluate. Also, perhaps because there has been more focus by researchers on economic aspects. In groundwater, there has not been much trading activity to have resulted in easily discernible impacts. In surface water, where there has been greater activity, environmental impacts do not appear to have been significant, but it should be noted that government has intervened by delivering water to important environmental sites. Moreover, water resources sustainability has been restored to some degree by means of buybacks of water entitlements from irrigators, albeit at great cost. Regarding social issues, water trading has had some positive impacts including the enabling of indebted farmers to service their debts by selling their entitlements. There have been indications, however, of disruption of community life and loss of local employment. There is also the perception that water markets, on their own, do not constitute a fair or acceptable process for allocating water.

Finally, Australia seems to be a good example of the increasing complexity of legislation in the management of water resources which, it may be argued, does not sit too well with the ideas of the free market. The fact that government considers it necessary to interfere to this extent reflects the difficulties of applying free market principles to a resource such as water.

England and Wales

Water resources

England and Wales are two of the constituent countries of the United Kingdom of Great Britain and Northern Ireland. Scotland is the third country and Northern Ireland the fourth, each having some devolved powers and legislation. England and Wales have a maritime temperate climate with an average annual rainfall of approximately 927 mm (1971–2000). Rainfall is higher in Wales, around 1,430 mm a^{-1} and lower in England, approximately 835 mm a^{-1}. It is driest in East Anglia, approximately 600 mm a^{-1}. In the last 40 years, there were periods of drought, when the average annual rainfall in England was more than 100 mm below the long-term average, notably the mid-1970s, 1988–91, 1995–97 and more recently 2004–06, (Met Office, n.d.). Effective rainfall (see Chapter 2 for definition) is about 50 per cent of rainfall (EA, 2008). It is lower in the east – in East Anglia it can be less than 200 mm a^{-1} – and higher in Wales and in southwest and northwest England. In southeast England it is generally greater than 400 mm a^{-1}.

Figure 6.8 Administrative map of England and Wales

The average annual total water abstraction (surface water (non-tidal) and groundwater) in England and Wales for the period 2000–2008 was approximately 13,570 Mm^3 (Defra, 2010). Groundwater comprised approximately 2,300 Mm^{3}a^{-1} or 17 per cent of the total. During the period, surface water abstractions (non-tidal) decreased from approximately 12,650 Mm^3 in 2000 to 10,520 Mm^3 in 2008. Groundwater abstraction fell only slightly from approximately 2,380 Mm^3 in 2000 to 2,140 Mm^3 in 2008. About 80 per cent of all groundwater abstraction is from central, south and east England and 9 per cent from northern England. The lowest abstraction is from Wales, less than 1 per cent. About three-quarters of groundwater is extracted from two main aquifers, both in England, namely the Chalk and the Permo-Triassic Sandstones. In 2003, abstraction from the Chalk was approximately 1,220 Mm^3 and from the Permo-Triassic Sandstones 555 Mm^3. Total recharge to the two aquifers was approximately 6,000 Mm^3 (EA, 2006). The total groundwater abstraction in the same year was approximately 2,390 Mm^3, 51 per cent from the Chalk and 23 per cent from the Permo-Triassic Sandstones.

In 2008, public water supply accounted for approximately 47 per cent of total abstraction, electricity generation just over 35 per cent, industry and fish farming 17 per cent, spray irrigation and farming about 0.6 per cent and other uses and private supply 0.3 per cent (Defra, 2010). Over the period, public supply consumption remained almost unchanged but water for electricity generation generally decreased by as much as 28 per cent in 2005 (Defra, 2010). In East Anglia, spray irrigation, although less than 1 per cent of average annual total use, can rise to about 20 per cent during a hot dry summer (EA, 2008). Groundwater provides much of the public water supply in England, ranging from more than 80 per cent in the centre, east and southeast to near 20 per cent in the southwest and north. In Wales, groundwater use is very low, less than 1 per cent.

Before 2004, there were approximately 46,400 abstraction licences (surface water and groundwater) in England and Wales. The number decreased to approximately 22,700 in 2005, as a result of the 2003 Water Act (see below) which did not require a licence for abstractions of 20 $m^3 d^{-1}$ or less. In 2008–09, there were 21,630 licences in force. However, the reduction in the number of licences did not significantly affect the licensed volume, which only decreased by 2–3 per cent. Approximately 24 per cent of all licences are in the Anglian region and 19 per cent in the Midlands. Approximately 8 per cent of all licences are for public water supply, 63 per cent for spray irrigation and agriculture, 2 per cent for electricity supply, 21 per cent for other industry and fish farming and 6 per cent for other uses (Defra, 2010). As indicated above, although, spray irrigation and agriculture hold the majority of abstraction licences, they account for less than 1 per cent of total abstraction. Actual abstraction is approximately 47–48 per cent of the licensed abstraction. Of this total, water companies abstract almost half, although 70 per cent is returned to the environment as sewage effluent (EA, 2008). Similarly, almost all water abstracted for electricity generation is returned.

The EA's Catchment Abstraction Management Strategies (CAMS) (EA, 2001) set out the resource availability status of England and Wales. Recent assessments suggest that groundwater is available for further abstraction in only a few areas, and overabstracted and overlicensed in many areas of East Anglia and the London Basin. For surface water and groundwater combined, available resources occur mainly in southwest and north and northwest England, and west Wales (EA, 2008).

Water legislation

Historically, water rights in England and Wales were governed by the riparian and absolute ownership doctrines. The licensing of water abstractions requiring volumes to be specified was not introduced until the middle of the last century (Water Resources Act, 1963). The Act required that water should be used beneficially. The issue of licences was by the competent water authority. Licences of right were issued to individuals or entities with statutory rights to abstract water, or to those who had abstracted water from a source at any time within a five-year period, before 1 April 1965. Protected rights were safeguarded from interference by new licences. In 1989, the ten water authorities established by previous legislation were privatised (Water Act, 1989). In 1991, an independent regulatory body, the National Rivers Authority (NRA), was established charged with a wide range of functions and duties over all aspects of water, including the granting of water abstraction licences. The role of the water authorities was restricted to providing services as water and sewerage undertakers. The NRA was replaced by the Environment Agency (EA) (Environment Act, 1995). Until 2003, all Acts maintained the land and water right link and transfer of water through succession. The 2003 Water Act separated land rights from water rights and allowed for water rights trading (Water Act, 2003, Chapter 37). The main aims of the Act were to achieve a sustainable use of water resources, strengthen the voice of consumers, bring a measured increase in competition, and promote water conservation. With regard to water rights, the 2003 Water Act provided as follows:

1   no abstraction licence required for quantities of 20 m^3 per day or less;
2   measurement of quantity as volume per unit time;
3   all new abstraction licences to be time limited with the provision for licences to be revoked or varied if not used for four years, and for the duration of a full abstraction licence of 12 years, though this may be reduced without compensation following a six-year notice period;
4   abstraction licences need not specify the land on which abstracted water can be used;
5   abstraction licences may be transferred by the holder of the licence to another person, with all transfers subject to approval by the EA;
6   all new abstraction licence applications must be accompanied by an environmental impact assessment report.

Stakeholders were given the opportunity to participate in decisions on the facilitation and management of water rights trading by responding to a consultation document circulated to 4,000 consultees in the period July–October 2003. There were only 70 responses, mostly from water companies and the agricultural sector. The process for the issue of abstraction licences is transparent involving the placement of notifications in the local press and the *London Gazette*.

Water rights transfers

The EA envisaged that water rights trading would ensure a more efficient use of water that would make surplus available for trade, and therefore allow other users to benefit in areas where water resources would otherwise not be available due to environmental constraints or full use of resources. In their booklet *A Guide to Water Rights Trading* (undated), the EA described the conditions under which the trading of water rights might take place. One of the main constraints related to the availability of water resources. As discussed above, in many areas of south and east England, groundwater is not available for further exploitation, which limits the scope of water rights transfers in these areas. Also, water rights trades may not be permitted in areas where water-dependent conservation sites need to be protected. Trading of groundwater rights is only allowed in the same aquifer (groundwater unit), and in surface water in the same river or between two tributaries in the same catchment (intrabasin).

The trading of water rights, whether in groundwater or surface water in England and Wales, has been extremely weak. In general, bought volumes were reduced by the EA to about three quarters of the existing licensed volumes (Harou, 2009), presumably on the basis of actual historic use but more likely in an effort to reduce abstraction. In the five-year period between July 2003 and August 2008, there were 48 licence trades. The total volume traded was 1.88 Mm³ or approximately 0.002 per cent of the total abstraction over the period. There was an increase in traded volumes from approximately 0.1 Mm³ a⁻¹ in 2004–2005 to an average of approximately 0.56 Mm³ a⁻¹ in 2006–2008. The majority of trades and volumes transferred were in East Anglia, followed by Thames and the Midlands. The lowest was in Wales (Table 6.10). There was very little water, only a small percentage of the total, transferred across different user sectors. Most water rights transfers, 52 per cent by volume, occurred within the spray irrigation/farming sector. There were only minor transfers between water companies, 5 per cent by volume, or from water companies to other users, 4 per cent by volume. Most transfers were temporary, 53 per cent by volume, whilst permanent transfers amounted to 24 per cent by volume.

In view of the weak water rights market, the EA and Ofwat (the Water Services Regulation Authority) undertook a study to review the barriers to water rights trading (EA/Ofwat, 2009). They made a number of proposals to facilitate trade, including:

1   an independent website dedicated to trading to facilitate the flow of information to the market;
2   the introduction of measures to liberalise the administrative process dealing with water rights trading;
3   the provision of further information to potential buyers and sellers on the nature and conditions of licences;
4   the introduction of measures to help achieve sustainable abstraction in the short and medium term, undertaken in a way such that it does not act as a disincentive to trading.

Table 6.10 Water trades in England and Wales (2003–2008) Trades by region (Data source: EA)

Region	Number of trades	Percentage by number (per cent)	Volume (Mm³)	Percentage by volume (per cent)
Anglian	31	66	1.003	53
Thames	5	10	0.429	23
Midlands	4	8	0.296	16
Southern	3	6	0.073	4
Southwest	2	4	0.024	1
North East	2	4	0.025	1
Wales	1	2	0.030	2
Totals	48	100	1.880	100

They suggested the following measures:

- reverse auctions or 'buy-backs' of unsustainable licences at the lowest price;
- ex gratia payments by the EA to abstractors;
- a review of the EA's abstraction licence structure to reflect the true economic, social and environmental value of water, including pricing signals to reduce abstraction; and
- legislative changes to enable the EA to make payments (buy-backs or ex gratia payments) to abstractors.

The EA/Ofwat proposals were considered further by Cave (2009) who was appointed by the Department of the Environment, Food and Rural Affairs (Defra) to undertake an independent review of competition and innovation in the water markets in England and Wales (Cave, 2009). The Cave Review endorsed the EA/Ofwat proposals for the introduction of legislation empowering the EA to undertake buy-backs (reverse auctions) of abstraction licences and for the increase of abstraction licence charges above cost recovery. It further suggested that a scarcity charge should be introduced in areas where licensed volumes remained at unsustainable levels. Also, where licence levels are sustainable, water rights to become fully tradable, subject only to modification for direct environment impacts and the impacts on other users from a change of use location. Ofwat (2010) in their report *Valuing water* noted that in view of the large share of water rights owned by water companies and of water abstracted (almost half of all the water abstracted) there were few opportunities for competition. In order to encourage new entrants to the water market, Ofwat (2010) proposed reforms to deregulate the 'upstream market' (this is the market which is concerned with water services that are upstream of retail of water to consumers, i.e. water resources, raw water distribution, water treatment and treated water distribution), but indicated that, initially, the greater scope for water trading was likely to be between appointed companies. The Colorado-Big Thompson project was cited as an example of the success of the market approach. As already discussed the project was funded by government and involved the conveyance of large amounts of water from a distant source in the Rockies, an area where there were ample water resources. Transfers across sectors were generally from low-value agricultural water to high-value urban water. A comparable scheme in England and Wales would be the transfer of water from Wales or from Scotland where there is abundant surface water to the drier east and southeast England, where the demand due to the higher population is greater.

In 2010–11, two water companies, Severn Trent Water and Anglian Water, entered the discussion on water rights trading. Ernst & Young/Severn Trent Water (2011) in their report *Changing course through water trading* considered that water trading across company boundaries from areas where water is available to areas of water scarcity utilising mainly the existing pipe network would benefit the environment and reduce upward pressure on bills to customers. They suggested that given the relatively small potential size of the market and cost, structural changes to the water sector should be kept to a minimum. They proposed six changes to develop water trading, including:

- harmonisation of costs for buyers and sellers;
- improving the flow and quality of information;
- the introduction in the water resources management plan of a requirement for water trading as an option to meet future demand; and
- unbundling the current combined water supply licence and creating a new upstream-only licence for new entrants.

Frontier Economics/AnglianWater (2011) in their report *A right to water?* drawing mainly from the Australian experience, reviewed four options for reform, similar to those identified by Cave (2009) and EA/Ofwat (2010):

1   improvement of administrative arrangements – it was suggested that in order to increase security for time-limited water rights holders there should be incorporated within legislation provision for their automatic renewal;
2   in order to protect the environment in conditions of severe scarcity there should be provision for buy-backs of water rights by government;
3   scarcity charges, the provision for which they considered to be a complex process that in isolation was not justified;
4   proportional reductions in water rights of all users in proportion to the relative share of the water resource. Such reductions, they considered to be inappropriate unless a strong and functional water market develops.

With regard to barriers to trade, they proposed increasing the visibility of the market, reducing transaction costs, and removing regulatory disincentives to trade between water companies. Also, they proposed reducing the uncertainty with regard to government future policy with regard to availability, which encourages water users to hold on to their water rights. In the light of the experience of water rights trading to date, the report suggested that active trading markets in England and Wales will be relatively small and discrete, and unlikely to resemble the scale of trading outcomes in other countries with water markets, where trading has been dominated by trade between agricultural users. The report noted, however, that one reason why a large water market has not developed was because there has been an absence of scarcity, which has been shown to be a major driver to trade in other countries.

Based on the Cave (2009) report, the government set out subsequently in a White Paper its case for the evolutionary reform of the water sector. This was followed by a Draft Water Bill in July 2012.

Comment on the water rights transfers in England and Wales and proposed reforms

In those countries where groundwater rights trading has been promoted, it has been with the objective of economic efficiency, which essentially meant the transfer of low-value, high-volume irrigation water to high-value, low-volume municipal supply and industrial use or for the growing of cash crops. Irrigated agriculture in these countries uses 70 per cent or more of total abstraction and public water supply generally 15–20 per cent. In England and Wales, the opposite is true. The volume of water used in agriculture, mainly for spray irrigation, is a mere fraction (less than 1 per cent) of total abstraction, whereas public supply and industry consume almost 90 per cent. Moreover, as water is sold at its economic value, there are no significant economic benefits of transferring water from agriculture to water supply or industry. This leaves very little scope for large groundwater (or surface water) rights transfers to other sectors. As a result, water rights markets in England and Wales are likely to remain small and localised (a view echoed by Frontier Economics/Anglian Water (2011)), and therefore, unlikely to have a significant impact on the management and sustainability of water resources. It has been suggested that trading would benefit if legislation allowed for the automatic renewal of abstraction licences, on their expiry. This is a rather retrogressive step, since time-limited licences were recently introduced in order to reduce overabstraction and ensure changes compatible with environmental sustainability. Buy-backs of abstraction licences by government, which in England and Wales are likely to be mainly from water companies which hold the bulk of the water rights volume, are not strictly 'free market' but rather government intervention to remove risks to the water market. In themselves, buy-backs do not create more water, and therefore, companies in addition to efficiency measures, may need to seek water elsewhere. Sharing of water with or without trades between water companies is a way forward, but availability of both surface water and groundwater, especially in the south and southeast of England is likely to prove a limiting factor. This was highlighted in a collaborative project of water sharing in East Anglia involving Anglian Water, Suffolk Water and Cambridge Water (Anglian Water/Cambridge Water Company and Essex & Suffolk Water, 2010). The long-term solution to satisfy all competing demands does not seem to lie in water rights trading in a small finite resource, but in undertaking large water transfer projects from areas of plenty, such as Wales or Scotland, to the areas of scarcity in the south and southeast of England. As already discussed in the previous chapter, there are many examples of such projects throughout the world, generally funded by government but recently also by the private sector.

Summary

The trading of water rights for both groundwater and surface water in England and Wales has been feeble, about 0.002 per cent of total abstraction. Reforms have been suggested to encourage trade, many borrowed from Australia, where as already discussed groundwater trading has been weak. It seems that the limiting factor to trading is the availability of resources, and despite any future reforms, the water rights market is likely to stay small and not able to provide a substantive solution to the allocation of resources.

Concluding remarks

The review of international experience indicates that there have been no market-led transfers of large quantities of groundwater, from low-value agricultural water to high-value urban water supply, using transferable water rights as the vehicle. Transferable groundwater rights were used, in conjunction with other measures, to reduce groundwater overabstraction in the Edwards Aquifer in Texas. However, this was achieved not through water rights trading but by the purchasing (and retiring) of groundwater rights from irrigators. Other measures included the provision of financial incentives to farmers to reduce abstraction and the buying of land in the recharge areas. This was a costly affair borne by increases in users' fees mainly from the non-agricultural sector, an unlikely scenario for many of the poor communities in developing countries. In Australia, it was felt necessary to buy back water access entitlements from irrigators at great cost to safeguard the environment during the recent drought.

The present review did not produce evidence to indicate that in the five countries examined water rights trading has been substantial or that it has contributed significantly to the sustainable management of water resources. In fact, water rights trades generally amounted to only a small percentage of total use. In Chile, where trading was allowed to proceed without interference from government, water rights were overallocated with groundwater being overabstracted in most areas, forcing government to introduce new legislation to protect aquifers and ecological flows. A similar situation exists in the western USA where many aquifers have been depleted, despite water rights trading having taken place for many years. In Mexico groundwater continues to be overexploited with the market being unable to provide substantive solutions. In Australia groundwater trading has been very weak, amounting to a few per cent of total groundwater use, and this despite considerable and costly administrative reforms over a number of years. In England and Wales, where unlike the other countries examined only less than 1 per cent of total use goes to irrigation, the trading of water rights has been too insignificant to make any difference in the allocation and management of water resources.

In general, there have not been many water transfers from irrigation to water supply, most transfers having taken place within the agricultural sector, and most having been temporary rather than permanent. There are only a few cases of poor farmers having adopted measures to conserve water and use it more efficiently so that they could trade the surplus for a higher price. In fact, poor farmers are usually sellers of water rights and seldom buyers. Overall, the experience has not been one of movement of substantial quantities from low-value irrigation groundwater to high-value municipal or industrial use. In Chile in addition to aquifer depletion, there have been environmental problems, problems of social equity, monopolies and hoarding of water rights.

In the countries examined, governments still pursue water rights trading as the favoured option for the efficient allocation of resources, and do so often at considerable expense. Although water rights trading may have a role to play, the ideal market of many buyers and many sellers has not materialised, and as a result, the role of government, both as an initiator and a regulator, and also as the paymaster for protecting the environment, rural communities and the vulnerable, is not likely to go away.

References

Chile

Arrueste, J. (2008) Water scarcity in Northern Chile for mining projects: present and future, 8th International Conference Water in Mining, Santiago, 9–11 July.

Bauer, C.J. (2004) *Siren song: Chilean water law as a model for international reform*. Washington DC: Resources for the Future.

Budds, J. (2004) Power, nature and neoliberalism: The political ecology of water in Chile, *Singapore Journal of Tropical Geography*, 25(3): 322–42.

Budds, J. (2009) Contested H_2O: Science, policy and politics in water resources management in Chile, *Geoforum*, 40(3): 418–30.

Budds, J. (2010) Water rights and indigenous groups in Chile's Atacama. In R. Boelens, D.H. Getches and A. Guerara-Gill (eds) *Out of the mainstream. Water rights, politics and identity*. London: Earthscan.

Dirección General de Aguas (DGA) (1999) *Politica nacional de recursos hidricos*. Santiago: Ministerio de Obras Publicas.

Donoso, G. (1999) Analisis del funcionamiento del mercado de los derechos de aprovechamiento de agua e i dentificacion de sus problemas. *Revista de Derecho Administrativo Economico* 1(2): 367–87.

Donoso, G. (2003) *Water markets: Case Study of Chile's 1981 Water Code*. Global Water Partnership South America. Pontificia Universidada Catolica de Chile. Santiago: Faculty of Agriculture and Forestry, Department of Agricultural Economics.

Dourojeanni, A. and Jouravlev, A. (1999). *El codigo de aguas de Chile: Entre la ideologia y la realidad*. Serie Recursos Naturales e Infraestructura 3, Division de Recursos Naturales e Infraestructura. Comision Economica Para America Latina. Santiago: United Nations.

Galaz, V. (2004) Stealing from the poor? Game theory and politics of water markets in Chile, *Environmental Politics*, 13(2) 414–37.

Gazmuri, R. and Rosegrant, M. (1994) Chilean water policy: The role of water rights, institutions, and markets. In M. Rosegrant and R. Gazmuri (eds) *Tradable water rights: Experiences in reforming allocation policy*. Arlington, VA: US Agency for International Development, Irrigation Support Project for Asia and the Near East.

Global Water Intelligence (GWI) (2010) A step in the right direction. 11(10). Available at http://www.globalwaterintel.com/archive/11/10/market-insight/step-right-direction.html

Hearne, R. (1995) The market allocation of natural resources: Transactions of water use rights in Chile. PhD dissertation, Department of Agricultural Economics, University of Minnesota.

Houston, J. and Hart, D. (2004) Theoretical head decay in closed basin aquifers: insight into fossil groundwater and recharge events in the Andes of northern Chile, *Quarterly Journal of Engineering Geology and Hydrogeology*, 37: 131–9.

Madaleno, I.M. (2007) The privatisation of water and its impact on settlement and traditional cultural practices in northern Chile, *Scottish Geographical Journal*, 123(3): 193–208.

Muñoz, J.F. (2003) Evaluation of groundwater availability and sustainable extraction rate for the Upper Santiago Valley Aquifer, Chile, *Hydrogeology Journal*, 11(6): 687–700.

National Water Commission (2005) *Australia water resources 2005: A baseline assessment of water resources for the National Water Initiative.* Canberra: Government of Australia.

Peña, H. (1996) Water markets in Chile: What are they, how they have worked and what needs to be done to strengthen them? Paper presented to the 4th Annual World Bank Conference on Environmentally Sustainable Development, 25–27 September, Washington, DC.

Rios, M. and Quiroz, J. (1995) *The market of water rights in Chile: Major issues.* Technical Paper No 285. Washington, DC: World Bank.

Romano, D. and Leporati, M. (2002) The distributive impact of the water market in Chile: A case study in Limari province 1981–1997, *Quarterly Journal of International Agriculture*, 41: 41–8.

Rosegrant, M. and Binswanger, H. (1994) Markets in tradable water rights: Potential for efficiency gains in developing country water resource allocation. *World Development* 22: 1613–25.

Rosegrant, M. and Gazmuri, R. (1994) *Reforming allocation policy through markets in tradable water rights: Lessons from Chile, Mexico, and California.* Environment and Protection Technology Division Discussion Paper No 6. Washington, DC: International Food Policy Research Institute.

Salazar, C. (2003) *Situacion de los recursos hidricos en Chile.* Mexico: Centro del Tercer Mundo para el Manejo del Agua, A.C.

Vergara, A. (1997a) El catastro publico de aguas: Consagracion legal, contenido y possibilidades de regulacion reglamentaria. *Revista de Derecho de Aguas* 8: 71–91.

Vergara, A. (1997b) Perfeccionamiento legal del mercado de derecho de aprovechamiento de aguas. Paper presented at fourth Convencion Nacional de Usuarios del Agua. 17–18 October, Arica, Chile. Santiago: Confederacion de Canalistas de Chile.

Vergara, A. (1997c) La libre transferibilidad de los derechos de agua: el caso Cileno. *Revista Chilena de Dercho*, 24(2): 369–95.

Vergara, A. (1998) *Derecho de aguas.* Santiago: Editorial Juridica de Chile.

World Bank (1994) *Peru: A user-based approach to water management and irrigation development.* World Bank Report No 13642-PE. Washington, DC: World Bank.

Mexico

Asad, M. and Garduño, H. (2005) *Water resources management in Mexico: The role of the Water Rights Adjustment Programme (WRAP) in water sustainability and rural development,* Sustainable Development Working Paper No 24. Washington, DC: The World Bank, Department of Latin America and Caribbean Region.

CONAGUA (2010) Statistics on water in Mexico, 2010 edition. Available at http://www.conagua.gob.mx/english07/publications/EAM2010Ingles_Baja.pdf

CONAGUA (2011) 2030 Water agenda. Available at http://www.conagua.gob.mx/english07/publications/2030_water_agenda.pdf

CONAGUA/OECD/IMTA (2010) Financing water resources management in Mexico, May 2010. Available at http://www.conagua.gob.mx/english07/publications/OECD.pdf.

Diario Official de la Federación (2009) Segunda Sección, Secretaria de Medio Ambiente y Recursos Naturales, Tomo DCLXX1 No 20, México, D.F.

Fortis, M and Ahlers, R. (1999a) *Naturaleza y extension del mercado de agua en el Distrito de Riego 017 de la Comarca Lagunera, Mexico*. Serie Latinoamericana: No. 10. Mexico: International Water Management Institute.

Fortis, M. and Ahlers, R. (1999b) *Mercado Lua en el Distrito de Riego 017: Compilation of research documents*. Mexico Country Program. Mexico: International Water Management Institute.

Garduño, H. (2005) Lessons for implementing water rights in Mexico. In B.R. Bruns, C. Ringler and R. Meinzer-Dick (eds) *Water rights reform: Lessons for institutional design*. Washington, DC: International Policy Institute.

Kloezen, W. H. (1998) Water markets between Mexican water user associations. *Water Policy* 1: 437–55.

Levina, E. (2006). *Domestic Policy Framework for Adaptation to Climate Change in the Water Sector Part II: Non Annex I Countries: Lessons Learned from Mexico, India, Argentina and Zimbabwe*, COM/ENV/EPOC/IEA/SLT (2006)11. Paris: OECD.

Tortajada, C. (2006) Water management in Mexico City metropolitan area, *International Journal of Water Resources*, Special issue: Water management for large cities 22(2): 353–76.

Wester, P., Minero, R.S. and Hoogesteger, J. (2011) Assessment of the development of aquifer management councils (COTAS) for sustainable groundwater management in Guanajuato, Mexico, *Hydrogeology Journal*, 19(4): 889–99.

Wilder, M. and Romero Lankao, P. (2006) Paradoxes of decentralisation: Neoliberal reforms and water institutions in Mexico. *World Development*, 34(11): 1977–95.

World Bank (2009) *Poverty and social impact analysis of groundwater overexploitation in Mexico*. The World Bank, Latin American Caribbean Region.

Western United States of America

Brewer, J., Glennon, R., Ker, A. and Libecap, G. (2008) Water markets in the west: prices, trading and contractual form, *Economic Inquiry* 46: 91–112.

Bryner, G. and Purcell, E. (2003) *Groundwater Law Sourcebook in the Western United States*. Boulder, CO: Natural Resources Law Centre, University of Colorado School of Law. http://www.colorado.edu/law/centers/nrlc/index.htm

Clifford, P., Landry, C. and Larsen-Hayden, A. (2004) *Analysis of water banks in the western states*. Washington, DC: Department of Ecology and Westward Research.

Contor, B.A. (2010) Status of groundwater banking in Idaho, *Journal of Contemporary Water Research & Education*, 144: 29–36.

Czetwertynski, M. (2002) *The sale and leasing of water rights in western states: An overview for the period 1990–2001*. Water Policy Working Paper No 2002 002. Atlanta, GA: Georgia State University.

Dellapenna, J. W. (2005) Markets from water: Time to put the myth to rest? *Journal of Water Research and Education*, 131: 33–41.

Dellapenna, J.W. (2012) The myth of markets for water. In J. Maestu (ed.) *Water trading and global water scarcity: International experiences*. Carbondale, IL: RFF Press/Routledge.

Donohue, Z. (2009) Property rights and western United States water markets, *The Australian Journal of Agricultural and Resource Economics*, 53: 85–103.

EAA (2006) *Comprehensive annual financial report for the fiscal year ended 12.31.06 and 12.31.05*. San Antonio, TX: EEA.

EAA (2010) *Groundwater management plan 2010–2015*. San Antonio, TX: EEA.

EAA (2012) 2012 Operating budget, adopted November 8, 2011. www.edwardsaquifer.org/files/FY2012_Budget_Document.pdf

Faunt, C.C. (ed.) (2009) *Groundwater availability of the Central Valley Aquifer, California,* US Geological Survey, Professional Paper 1766. Reston, VA: US Geological Survey.

Hanak, E. (2003) *Who should be allowed to sell water in California. Third party issues and the water market.* San Francisco, CA: The Public Policy Institute of California, San Francisco.

Howitt, R. E. (1994) Empirical analysis of water market institutions: The 1991 California water market. *Resource and Energy Economics,* 16(4): 357–71.

Howitt, R. and Sunding, D. (2004) Water infrastructure and water allocation in California. In J. Siebert (ed.) *California agriculture: Dimensions and issues.* Berkeley, CA: University of California Giannini Foundation of Agricultural Economics and Division of Agricultural and Natural Resources.

Jarvis, W. T. (2011) Unitisation: a lesson in collective action from the oil industry for aquifer governance, Special issue: Strengthening cooperation on transboundary groundwater resources, *Water International,* 36(5): 619–30.

Kenny, J.F., Barber, N.L., Hutson, S.S., Linsey, K.S., Lovelace, J.K. and Maupin, M.A. (2009) *Estimated use of water in the United States in 2005,* US Geological Survey Circular 1344. Reston, VA: US Geological Survey.

Libecab, G. D. (2010) Water rights and markets in the US semi-arid west: Efficiency and equity issues. 'The Evolution of Property Rights Related to Land and Natural Resources' conference, September 20–21, Lincoln House, Cambridge, MA.

McGuire, V.L. (2007) *Water-level changes in the High Plains Aquifer, pre-development to 2005 and 2003 to 2005,* US Geological Survey Scientific Investigations Report 2006-5324. Reston, VA: US Geological Survey.

Maupin, M.A. and Barber, N.L. (2005) *Estimated withdrawals from principal aquifers in the United States, 2000,* US Geological Survey Circular 1279. Reston, VA: US Geological Survey.

Semitropic Water Storage District (n.d.) Monitoring Committee. http://www.semitropic.com/MonitoringComm.htm

Steinhart, P. (1990) The water profiteers, *Audubon* March: 38–51.

USBR (2007) *Semitropic Stored Recovery Unit special study report.* Sacramento, CA: US Bureau of Reclamation, Mid Pacific Region.

USGS (n.d.) National Atlas of United States of America, Annual Precipitation 1961–1990, http://www.nationalatlas.gov/mld/prism0p.html

Australia

ABS (2008) *Water and the Murray–Darling Basin: statistical profile, 2000–01 to 2005–06.* Canberra: Australian Bureau of Statistics. Available at http://www.abs.gov.au/ausstats/abs@.nsf/Latestproducts/C21D2B9AF8C6C309CA2574A5001F6CB8?opendocument

ABS (2010) *Water account Australia 2008–09.* Canberra: Australian Bureau of Statistics. http://www.ausstats.abs.gov.au/Ausstats/subscriber.nsf/0/D2335EFFE939C9BCCA2577E700158B1C/$File/46100_2008-09.pdf

Bjornlund, H. (2004) Formal and informal water markets: Drivers of sustainable rural communities? *Water Resources Research,* 40(9): W09S07.

Bjornlund, H., Wheeler, S. and Rossini, P. (2012 Water markets and their environmental, social and economic impacts in Australia. In J. Maestu (ed.) *Water trading and global water scarcity: International experiences.* Carbondale, IL: RFF Press/Routledge.

Bureau of Meteorology (n.d.) Current month to date rainfall totals for Australia http://www.bom.gov.au/jsp/ncc/climate_averages/rainfall/index.jsp

Bureau of Meteorology (2011) *Australian water resources assessment 2010: Summary Report.* Canberra: Australian Government.

COAG (1994) Water Reform Framework: Hobart, Communiqué 25 February, 1994.

COAG (2003) Communiqué 29 August, 2003, http://archive.coag.gov.au/coag_meeting_outcomes/2003-08-29/index.cfm

COAG (2004) Communiqué 25 June, 2004, http://archive.coag.gov.au/coag_meeting_outcomes/2004-06-25/index.cfm

Crase, L., O'Keefe, S. and Dollery, B. (2009) Water buy-back in Australia: Political, technical and allocative challenges. 53rd Annual Australian Agricultural and Resource Economics Society Conference, February 11–13. Cairns, Australia.

Edwards, J., Cheers, B. and Bjornlund, H. (2008) Social, economic and community impacts of water markets in Australia's Murray–Darling Basin region, *International Journal of Interdisciplinary Social Sciences*, 2(6): 1–10.

Fenton, M. (2006) *A survey of beliefs about permanent water trading and community involvement in NRM in the Loddon Campaspe irrigation region of Northern Victoria*, Report to the North Central Catchment Management Authority, Victoria. Melbourne: Department of Primary Industries.

Frontier Economics (2007) *The economic and social impacts of water trading: Case studies in the Victorian Murray Valley*, Report for the Rural Industries Research and Development Corporation, National Water Commission and Murray–Darling Basin Commission, RIRDC Publication No. 07/121. Canberra: RIRDC.

Godden, L. (2005) Water law reform in Australia and South Africa: sustainability, efficiency and social justice, *Journal of Environmental* Law, 17(2): 181–205.

McKay, J. (2011) The legal frameworks of Australian water: Progression from common law rights to sustainable shares. In L. Crase (ed.) *Water policy in Australia: The impact of change and uncertainty*. Washington, DC: RFF Press.

Mitchell, M., Curtis, A., Sharp, E., and Mendham, E. (2011) *Social research to improve groundwater governance: literature review*, ILWS Report No. 66. Albury, NSW: Institute for Land, Water and Society, Charles Sturt University.

Murray–Darling Basin Authority (MDBA) (2011) *Proposed basin plan*. Report for Murray–Darling Basin Authority. Canberra: MDBA.

Murray–Darling Basin Commission (2008) *Risks to shared water resources, Groundwater status report 2000–05*. Prepared by URS for the MDBC. Canberra: MDBC.

Musgrave, W. (2011) Historical development of water resources in Australia: Irrigation policy in the Murray–Darling Basin. In L. Crase (ed.) *Water policy in Australia: The impact of change and uncertainty*. Washington, DC: RFF Press.

NLWRA (2001) Australian Water Resources Assessment 2000. Surface water and groundwater – availability and quality, *Australian Natural Resources Atlas*, National Land and Water Resources Audit. www.anra.au/topics/water/pubs/national/water_contents.html

NWC (2007) *Australian water resources 2005: A baseline assessment of water resources for the National Water Institute key findings of the Level 2 Assessment Summary Results*. Canberra: National Water Commission.

NWC (2010) *The impacts of water trading in the southern Murray–Darling Basin: an economic, social and environmental assessment*. Canberra: National Water Commission.

NWC (2011a) *Water markets in Australia: a short history*. Canberra: National Water Commission.

NWC (2011b) *Australian water markets: trends and drivers 2007 to 2010–11*. Canberra: National Water Commission.

NWC (2012) *Impacts of water trading in the southern Murray–Darling Basin between 2006–07 and 2010–11*. Canberra: National Water Commission.

Productivity Commission (2006) *Rural water use and the environment: The role of market mechanisms.* Research Report. Melbourne: Productivity Commission.

Quiggin, J. (2011) Uncertainty, risk and water management in Australia. In L. Crase (ed.) *Water policy in Australia: The impact of change and uncertainty.* Washington, DC: RFF Press.

Sinclair Knight Merz (2008) *Great Artesian Basin sustainability initiative: Mid-term review of Phase 2.* Canberra: Australian Government, Department of the Environment and Water Resources.

URS (2010) *Murray–Darling Basin return flows investigation: final report,* prepared for the Murray–Darling Basin Authority. Southbank, VIC: URS Australia.

England and Wales

Anglian Water/Cambridge Water Company and Essex & Suffolk Water (2010) *Trading theory for practice.* Huntingdon: Anglian Water. http://www.anglianwater.co.uk/about-us/statutory-reports/B439A384462B42ADB129C941BD122448.aspx

Cave, M. (2009) *Independent review of competition and innovation in water markets: Final report.* London: Defra.

DEFRA (2010) *Estimated licensed and actual abstractions from all surface and groundwater sources 1990–2008,* e-digest of Environmental Statistics. London: Environmental Statistics Service, DEFRA.

EA (2001) *Managing water abstraction: The catchment management abstraction strategy process.* Bristol: Environment Agency.

EA (2006) *The state of groundwater in England and Wales.* Bristol: Environment Agency.

EA (2008) *Water resources in England and Wales – current state and future pressures.* Bristol: Environment Agency.

EA (undated) *A guide to water rights trading.* Bristol: Environment Agency.

EA/Ofwat (2009) *Review of barriers to water rights trading: Final report.* Bristol: Environment Agency/Ofwat Water Rights Trading Project.

Ernst & Young/Severn Trent Water (2011) *Changing course through water trading.* Bringham: Severn Trent http://www.stwater.co.uk/upload/pdf/STW_Water_Trading_FINAL_9_June_2011.pdf

Frontier Economics/Anglian Water (2011) *A right to water? Meeting the challenge of sustainable water allocation. Main report.* Huntingdon: Anglian Water. http://www.anglianwater.co.uk/_assets/media/a-right-to-water-full-report.pdf

Harou, J. (2009) Promoting water trading in England and Wales, presentation at Rethinking Water: Emerging challenges for Regulation & Legislation, CIWEM conference, December 2009, London.

Met Office (n.d.) UK rainfall. http://www/metoffice.gov.uk/climate/datasets/Rainfall

Ofwat (2010) *Valuing water. How upstream markets could deliver for consumers and the environment.* Birmingham: Ofwat. http://www.ofwat.gov.uk/publications/focusreports/prs_inf_value.pdf

7 Guidelines for the introduction of transferable groundwater rights (TGR)

This chapter is concerned with guidelines for the introduction of transferable groundwater rights (TGR). It begins with a brief discussion of the reasons why guidelines are needed and sets out a list of general guidelines. A transferable groundwater rights evaluation model is presented together with a detailed discussion of the key factors that are likely to influence the evaluation process. The chapter concludes with the example of the Arani-Kortailayar basin aquifer in India, which illustrates how some of the key factors have been addressed.

Need for guidelines

The review of international experience in Chapter 6 has indicated that the introduction of TGR is a complex process. In addition to economic considerations, it has to take account of aspects of resource sustainability, environmental protection, social equity, and sensitivity to cultural and political perceptions. Strong political ideology has been a factor – and often the driving force – behind the adoption of market principles for the management of water resources. This has not always allowed arguments to evolve in an objective manner. For these reasons, a set of guidelines that address all relevant issues systematically should be useful. It is hoped that the transferable groundwater rights evaluation model (TGREM) described below should go some way in providing a systematic approach to the decision process.

General guidelines

In considering the introduction of TGR, it is useful to bear in mind the following general points:

1　a 'top-down' approach should be avoided;
2　consultations with stakeholders are essential;
3　transparency is important in order to avoid suspicion that there are underlying motives – this should include the dissemination of information in easily understandable formats;

4   implementation of reforms should be in stages – a staged approach allows stakeholders to see and evaluate the benefits of reforms and for changes to be made in the light of experience;

5   reforms take a long time to implement.

Experience shows that reforms, especially if controversial, imposed from above often run the risk of failure, take too long to implement or become so diluted that they fail to meet their original objectives. It is important therefore that the concepts, objectives, benefits and risks are clearly explained, and that stakeholders and others are consulted for their views and allowed to express their concerns. On a local level, this is best done in meetings and workshops, especially in developing countries where poor farmers may not have the facilities or knowledge otherwise to acquire information. Educational material and dissemination of information in the local and national press, radio, television and the internet can reach a wider group of people. Phone-in discussions on radio and television debates will inform both the public and interested parties on the issues, and highlight advantages and disadvantages. As TGR may involve trading, their introduction may be seen as privatisation 'through the back-door'. A transparent approach should allay suspicions of hidden motives. As must have become apparent in earlier chapters, water governance is complex and multifaceted. A staged approach should address more easily any issues that may arise in the course of introducing TGR and enable changes to be made in the light of experience. Reforms cannot be implemented overnight and should be allowed to take their time. A hasty approach may in fact prolong rather than shorten the process. Experience shows that changes may take many years to implement. A good example is the Edwards aquifer where legislation has taken about 15 years to implement, and even recently changes in the water rights ownership regime have been challenged. In Chile and Mexico, where new water codes were enacted 20 to 30 years ago, not all water rights have been registered. In both these countries, and in Australia, legislation is still being added to and reformed to take account of physical conditions affecting water resources, sluggishness in water rights trades, and environmental and social concerns. In England and Wales, approximately ten years after allowing water rights to be transferable, the level of trades has been miniscule, and further legislative reforms are being considered.

Transferable groundwater rights evaluation model (TGREM)

Given that the introduction of TGR can be a complicated process that involves the interplay of diverse factors, a generic flowchart type model such as the one shown in Figure 7.1 may assist in the decision-making process. The model consists of a step-by-step logical process of questions, answers and actions. It may be added to or modified to suit particular conditions or jurisdictions, as the case may be.

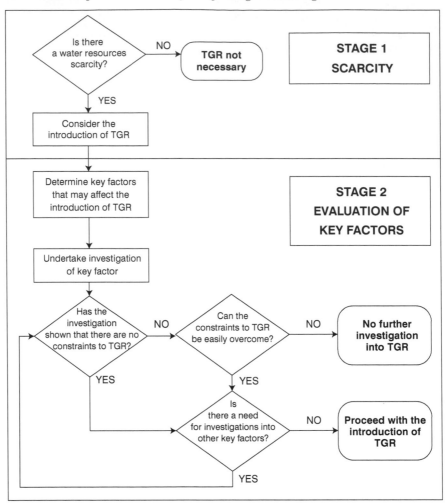

Figure 7.1 Transferable groundwater rights evaluation model (TGREM)

The model examines the introduction of transferable groundwater rights in two stages. Stage 1 investigates scarcity of water resources. Stage 2 evaluates the key factors that may affect the introduction of transferable groundwater rights. By means of an iterative process of evaluating each key factor a conclusion is reached on whether or not transferable groundwater rights should be introduced.

Water resources scarcity is the main factor that underlies the concept of TGR. Without scarcity there is no need for TGR. If Stage 1 indicates scarcity, Stage 2 may follow. The first task to be undertaken in Stage 2 is to identify the key factors that may affect the introduction of TGR. Typical key factors may include physical aspects of aquifers, legislation, economics, politics, the environment and social and cultural aspects. The number of key factors may vary depending on the local conditions of the country or jurisdiction. Identifying the key factors always carries the risk of bias, which may be minimised if it is undertaken in consultation with representatives of all main stakeholders. Unavoidably, this may result in a protracted process.

The flowchart logic predicates that the investigation of each successive key factor depends on the outcome of the investigation of the previous one. Thus, following the investigation of a particular key factor, if the answer to the question of whether there are constraints to the introduction of TGR is a 'no' the process continues with the evaluation of the next key factor, should there be one. If the answer is a 'yes', the question is asked whether the constraints could be easily overcome. If the answer to this question is a 'yes' the process continues with the evaluation of the next key factor, should there be one. A 'no' answer means that the process stops and no further consideration of TGR takes place. For a positive conclusion to be reached, i.e. one that recommends the introduction of TGR, the evaluation should conclude that either there are no constraints or any constraints that have been identified may be overcome.

Groundwater resources scarcity

From a water resources viewpoint, in renewable aquifers scarcity occurs when over a period, usually of a few years, abstraction exceeds replenishment. Indications of scarcity may be of a qualitative nature or anecdotal (falling groundwater levels, reductions in the discharge of boreholes and springs, wells drying up, ecological changes in wetlands) and may need to be confirmed by investigation. As groundwater is not a visible resource, hearsay reports can, at times, be misleading. But even with quantitative investigations, it may take many years of monitoring and observation to establish whether groundwater is actually being permanently depleted. Investigations should aim at determining the sustainable yield of the groundwater resources, which must be used as the basis for allocating water to different users, including the environment. The scope of likely investigations to determine groundwater scarcity may include:

1   hydrogeological and hydrological studies, monitoring and the use of groundwater computer models to assess aquifer recharge, water balance and sustainable yield;
2   review of impacts of groundwater abstraction on water quantity, water quality, river flows and the environment;
3   surveys to determine the volumes of water abstracted for different uses, the number of abstractors, and the number of wells/boreholes;
4   estimation of future water demand for different sectors.

Key factors and constraints

Aquifer connectivity

As has already been discussed in Chapter 6, groundwater rights trading in Australia has apparently been hampered by the fact that groundwater bodies are discrete and not hydraulically connected. In Chile the lack of hydrologic connectivity has been a constraint in the development of a surface water rights market across basins. Many alluvial or fractured basement rock aquifers have a limited areal extent and can only be developed locally. Similarly, large aquifer systems may lack hydraulic continuity over long distances. This can pose obstacles to water transfers which may impede groundwater rights trading, unless engineering schemes involving conveyor pipelines or canals are implemented. Some alluvial aquifers are thin and have a small carry-over storage. Therefore, they may not always be able to provide a reliable supply for municipal use, especially during periods of drought when water is needed most. Developing such aquifers for large urban supplies may require a large number of wells, which may be impractical and uneconomic.

Legislation

The basis of TGR is that groundwater rights can be transferred independently of land. Groundwater rights may thus be treated as property rights that can be freely traded as a commodity. There has been a general reluctance to treat water in this manner. In most countries, groundwater rights are still linked to land and legislation is needed to separate the two. A constitutional requirement for public ownership of water resources is not necessarily an obstacle to water rights being treated as property rights. Indeed, even in Chile where water rights have been assigned clear property right status, water is recognised as a national asset for public use, in all its forms and wherever it is found. As discussed in Chapter 3, modern water codes treat water rights as user rights, which allow the use of water resources by private individuals or entities while the ownership of water resources remains vested in the state. The land–water link is a particularly serious constraint to TGR, as a water right cannot be traded on its own. This can only be overcome by legislation that divorces water rights from land ownership. However, it may be necessary for conditions to be imposed on groundwater rights transfers in order to safeguard buyers and sellers, protect aquifers from depletion and prevent significant social and environmental effects. Of course, regulation itself usually presents an obstacle to trading and may deter prospective participants from entering the water rights market. Poorly defined groundwater rights may have the same effect. Thus, water rights reforms may need to be implemented before the introduction of TGR.

Economics

A successful economic model requires a well-developed market of many sellers and many buyers. Experience so far suggests that water markets have been limited. The primary objective of TGR is seen by economists as achieving a better economic value for water and an economically efficient allocation of groundwater resources. Constraints preventing the achievement of economic objectives may include: poorly defined and/or insecure groundwater rights, high transaction costs, lack of opportunity costs, inadequate infrastructure to convey water and poor aquifer connectivity. However, it seems doubtful whether the ideal market could be achieved, even if most of the constraints were to be removed. The most that can be hoped for is an increasing awareness of the economic value of water and therefore a more sustainable utilisation of groundwater resources.

Politics

As already mentioned, TGR and water rights trading has been largely driven by political ideology. Politics drives legislative change, and depending on who governs, it can move swiftly, lie dormant for many years or never be implemented. Water rights trading is an emotive issue and soon attracts the attention of politicians, NGOs and pressure groups. In a democratic country, an understanding of their concerns and strength of support or opposition to TGR is therefore essential.

Environment

TGR may be used by government for buy-backs of groundwater rights, as has been done in the Edwards Aquifer, Texas to prevent damaging levels of abstraction. However, impacts to the hydro-environment are an added risk to groundwater rights trading, which makes potential buyers apprehensive, unless government is willing to intervene with measures to mitigate the risk. When water from irrigated agriculture is moved to urban supply, land that has been left uncultivated may be affected by soil erosion. In dry climates or during long summers, dust may become a serious health problem.

Social aspects

There are good ethical and moral reasons, but also aspects of social stability and cohesion, that require consideration of the impact of TGR and water rights trading on rural communities in some developing countries. Social issues concern the weakest in society, who have neither the power nor the means to confront changes that may affect adversely their livelihoods and welfare. Poor farming communities, indigenous populations, nomadic and pastoral societies and those who are disadvantaged in caste social structures fall into this category. Rural communities almost invariably rely on groundwater as their main source of water for drinking, washing and watering small plots of land for food. Poor farmers may be enticed with offers of relatively large sums of money by rich landowners, municipalities, agricultural businesses or mining corporations to sell their groundwater rights. Without water many are not be able to grow their own food and may be forced to move into towns and cities in search of employment. Poor people are invariably sellers of groundwater rights and seldom buyers. This can result in the concentration of groundwater rights in the hands of the few richer members of society, which may disadvantage the less fortunate.

Cultural aspects

Cultural aspects relate to local traditions, customs and religious beliefs. In some cultures, water is considered to be a god-given good to be freely enjoyed by all and not to be traded for profit. While in others, there is a strong belief that groundwater is inextricably linked with land, and the landowner has a right to it that cannot be challenged. Custom and tradition underlies the rights of nomadic and pastoral people to water sources, for themselves and their livestock.

Investigations of key factors

Table 7.1 presents a brief description of the objectives and scope of investigations that need to be considered to determine the relevance or importance of each key factor to the introduction of TGR.

Table 7.1 Investigation of key factors to be considered in the introduction of TGR

Aquifer connectivity

Investigation objectives: To determine whether aquifer connectivity can affect the transfer and trading of groundwater rights.

Suggested scope of activities

- examine the location and extent of aquifers in relation to proposes uses;
- examine the hydraulic connectivity of aquifer systems and basins;
- consider constraints to groundwater transfers;
- review and cost measures to enable the hydraulic connectivity between groundwater sources and areas of supply.

Legislation

Investigation objectives: To determine: (a) whether TGR are compatible with or permitted by the existing or near future national and state legislation; (b)required legislative changes to include TGR in water codes.

Suggested scope of activities

- identify constitutional position with regard to water resources ownership;
- review and analyse existing legislative and institutional systems in relation to groundwater rights;
- review and analyse relevant court cases;
- review the relationship between water rights and land ownership;
- review informal (pastoral, customary or communal) water rights;
- determine whether or not the existing or any near future planned legislation explicitly prohibits TGR;
- consider whether there is need to introduce formal, explicit and clearly defined groundwater rights;
- consider how legislation may be changed to allow for the introduction of TGR.

Economics

Investigation objectives: To determine whether TGR in a free market can:

1. regulate the use of groundwater to a sustainable level;
2. lead to the efficient use and allocation of water resources;
3. enable water to achieve its economic value.

Suggested scope of activities

- review the existing allocation of water and whether there are significant differences in the economic value of water in different sectors, consider whether water rights trading can have a significant effect on water achieving its economic value;
- undertake willingness-to-pay and ability-to-pay surveys;
- undertake surveys to determine the total economic value (TEV) of water;
- consider how water markets can prevent overabstraction and how they can protect water resources and the environment from undesirable consequences, refer to relevant experiences elsewhere;
- examine the implications of reallocating water from low-value uses (irrigated agriculture) to high-value uses (urban supply, industry, mining, cash crops);
- review subsidies to agriculture and other sectors, if applicable;
- consider the economic impacts on the poor, and whether they are likely to be buyers or sellers of water rights;
- consider the risk and impact of monopolies;
- consider the impact of loss of water rights on land prices;
- consider the economic impact on the environment;
- by taking account of all relevant factors estimate the economic value and economic cost of water;
- consider what economic reforms may be needed and the cost of implementing them in order to overcome or mitigate barriers to water rights trading;
- estimate the likely volume of trades in relation to total abstraction;
- conclude whether and under what conditions TGR and water rights trading may be able to achieve groundwater resources sustainability, economic efficiency and economic value for water.

Politics

Investigation objectives: To determine whether the political climate supports TGR.

Suggested scope of activities

- identify whether the political climate favours a free market on water resources;
- investigate the views of local politicians, and their strength of support or opposition to TGR, do the same with non-governmental organisations (NGOs), local communities and pressure groups.
- discuss whether politics are likely to pose serious obstacles to the introduction of TGR, and consider whether these could be overcome.

Environment

Investigation objectives: To determine whether TGR are likely to have an adverse effects on the environment.

Suggested scope of activities

- determine the hydraulic relationships between aquifers, surface water bodies, and ecosystems;
- determine the extent to which groundwater flows and springs sustain wetlands, rivers, and the hydro-ecology.;
- identify current activities causing adverse effects;
- identify potential impacts on land and soils where irrigated land is to be made fallow following a change of groundwater use from irrigation to municipal/industrial use;
- examine whether and how TGR can prevent hydrological and ecological impacts;
- discuss whether TGR can help in protecting the environment. Consider what regulatory or other measures may be taken to mitigate or overcome any adverse effects that may result from the introduction of TGR and water rights trading.

continued ...

Table 7.1 continued

Social

Investigation objectives: Determining whether TGR are likely to have adverse social impacts.

Suggested scope of activities

- evaluate the social and economic impacts on poor farmers, indigenous populations and other vulnerable communities that are likely to result from selling their groundwater rights;
- consider whether groundwater rights belonging to such communities should be transferable and tradable;
- consider what measures (regulatory, financial or other) might be needed to mitigate or overcome adverse impacts.

Cultural

Investigation objectives: Determining whether TGR are compatible with local traditions, customs and religious beliefs.

Suggested scope of activities

- undertake surveys and studies to determine local customs, traditions and religious beliefs in relation to water ownership, use and tradability;
- evaluate whether TGR are compatible with local customs, traditions and religious beliefs;
- if not, consider whether appropriate regulatory or other measures may be possible to enable the introduction of TGR and water rights trading.

A case study for the introduction of TGR in the Arani-Kortalaiyar (A-K) groundwater basin, South India

The study was initiated by the World Bank and the Chennai Metropolitan Water Supply and Sewerage Board (CMWSSB) in view of the excessive groundwater abstractions for irrigation, about 95 per cent of total abstraction, which led to a steady depletion of the A-K basin aquifer and the dwindling of the yield of municipal wellfields supplying the city of Chennai (previously Madras). The study was carried out by Scott Wilson Piésold *et al.* (2004, 2006) in two phases. The first phase (2001–2004) involved extensive hydrogeological, hydrochemical, geophysical and modelling investigations to reassess aquifer sustainable yield. The second phase (2005–2006) considered the introduction of a system of formal, explicit and transferable groundwater rights. The objectives were, firstly to regulate the uncontrolled abstraction of groundwater and eventually reverse the undesired effects on quantity and quality and secondly, to enable the bulk transfer of groundwater from low-value agricultural use to higher-value urban supply.

Groundwater resources

The A-K groundwater basin is located in the state of Tamil Nadu, close to the city of Chennai. It has an area of approximately 1,500 km^2 and is bounded to the east by the Bay of Bengal. In the west it extends into the hills of Andhra Pradesh.

Rainfall is seasonal, during June–September and October–December. Long-term average annual rainfall is about 1,080 mm, of which 90 per cent falls during the monsoon periods. Land use is predominantly agricultural. Farmers grow three main crops. Rice paddy is the main crop followed by groundnut and pulses. In January–April crops are wholly sustained by irrigation, and in the rest of the year by rainfall supplemented by irrigation. When the monsoons fail groundwater is the main source of irrigation water.

The aquifer consists of sands and gravels, deposited in the past by two main rivers, the Arani and Kortalaiyar, which run from west to east. It is best developed in a buried channel, about 290 km^2 in area, which follows the old river courses. The aquifer is thin, ranging from a few metres to about 30 m and has limited storage. Because of this, the year to year carry-over capacity is small and the aquifer quickly becomes dewatered. Groundwater resources are, therefore, sensitive to recharge. Failure of the monsoon rains, from the late 1990s to 2004, combined with large uncontrolled irrigation abstraction, led to serious water shortages. In the coastal areas there was contamination of groundwater by seawater intrusion.

There are three main sources of groundwater recharge: rainfall infiltration (direct recharge), river bed infiltration (indirect recharge) and irrigation return flows. There is also a minor contribution from seepage through surface water reservoirs. The average total recharge for the period 1970–2002 was estimated to be 350 Mm^3 a^{-1} representing approximately 13 per cent of the rainfall over the period. Irrigation returns were a significant component, with approximately 40 per cent of the applied irrigation water returning to the aquifer. The relationships between recharge and abstraction were investigated by means of a finite element groundwater model (Charalambous and Garratt, 2009). Water abstracted for municipal supply is exported from the basin, and groundwater recharge relies mainly on rainfall and streambed infiltration. It is in the range of 80–140 Mm^3 a^{-1}, reaching a low of about 50 Mm^3 a^{-1} during low rainfall years. With irrigation, recharge becomes much higher, 220–280 Mm^3 a^{-11}, decreasing to around 150 Mm^3 a^{-1} during dry years. Without abstraction the aquifer soon fills up (it takes approximately 10 years from its depleted state) and recharge gradually decreases to approximately 10 Mm^3 a^{-1}. Abstraction creates space for recharge to enter.

Aquifer sustainable yield and scarcity

There are two competing uses for the groundwater of the A-K basin aquifer: for local irrigation, mainly for rice paddy, and for the municipal supply of Chennai. Abstraction for irrigation has been on average more than 400 Mm^3 a^{-1} from some 40,000 small-capacity, narrow-diameter boreholes, and for municipal supply approximately 20 Mm^3 a^{-1} from five wellfields situated 10–30 km from Chennai. It was originally planned to extract about 64 Mm^3 a^{-1} for the water supply of Chennai (UNDP, 1985 and 1987). However, a steady dewatering of the aquifer especially after 1990 and during 2000–2003, caused the yield of many of the municipal water supply boreholes to decrease and the quality of the groundwater of the boreholes closer to the coast to deteriorate.

Aquifer connectivity

The A-K basin alluvial aquifer although fairly extensive is mainly developed in the main palaeoriver buried channels. There is probably little hydraulic connection between the Arani and Kortalaiyar river valley deposits, except perhaps in the coastal area. Also, the aquifer rests on an undulating surface of low-permeability Gondwana shales and sandstones, which interrupts its hydraulic continuity. Outside the buried channel and in the upper reaches of the basin, the aquifer remains fairly thin, and, thus, not able to provide sustainable all-year-round abstractions, especially during periods of low rainfall. Because irrigation groundwater is used at or close to the point of abstraction, aquifer connectivity poses no serious problems to its exploitation. Moreover, the use of a large number of low-capacity boreholes enables abstraction to continue even when a substantial proportion of the aquifer has been dewatered (Charalambous and Garratt, 2009). For municipal supplies to Chennai, the economically efficient approach is to develop wellfields comprising relatively few high-capacity boreholes where the aquifer is most productive. This limits exploitation to a smaller part of the aquifer, which needs to remain substantially saturated if borehole yields are to be sustained. Transferring water from other less productive areas using clusters of low-capacity boreholes requires connecting pipeline networks and a higher level of maintenance and pumping costs.

Legislative aspects

As with other natural resources, water rights in India are governed both by customary practices and common law traditions on the one hand, and the sovereign right of state over natural resources on the other. The continuation of customary rights is implicit in Article 372 (1) of the Constitution of India, which states that:

> all the law in force in the territory of India immediately before the commencement of this Constitution shall continue in force therein until altered or repealed or amended by a competent Legislature or other competent authority.

With regard to principles of policy to be followed by the state article 39 requires that the state directs its policy towards securing:

> that the ownership and control of material resources of the community are so distributed as best to subserve the common good [and] that the operation of the economic system does not result in the concentration of wealth and means of production to the common detriment.

The role of the state as a public trustee has been used by the Supreme Court in the case of *M C Mehta* v. *Kamal Nath and others* 1997. The Supreme Court applied the Public Trust doctrine that issues from Articles 39 of the Constitution and held that the government as the functional arm of the state holds the natural resources such as land, air and water in trust for the people.

Based on English common law, groundwater rights in the Indian legal system have been linked to land ownership. A private landowner has an exclusive right to draw and use water under his land to an unlimited extent, even if this has the effect of depriving his or her neighbour of water (absolute ownership doctrine). This doctrine has been challenged by the Kerala High Court in *Perumatty Grama Panchayat* v. *State of Kerala* (2003.12.16) known as the Coca-Cola case. The council for the Coca Cola company maintained that in the absence of specific legislation prohibiting the extraction of groundwater, the company was free to extract any amount of groundwater which is available underground in the land owned by it. The court held that the principles that were applied in previous decisions relying on English decisions of the nineteenth century could no longer apply in view of the sophisticated methods for groundwater extraction and the emerging environmental jurisprudence. The court, invoking the doctrine of public trust, upheld the action of the Panchayat to prevent the company from extracting groundwater. This decision, although useful in itself, does not offer a comprehensive solution to groundwater overabstraction in India as a whole. Consequently, a number of states (Andra Pradesh, Goa and Kerala) have recently enacted groundwater legislation.

In 1987, the state of Tamil Nadu promulgated the Madras Metropolitan Area Groundwater (Regulation) Act (amended by the Act of 2002) in order 'to regulate and control the extraction, use or transport of groundwater and to conserve groundwater in certain areas in the State of Tamil Nadu'. The main purpose of the Act was to ensure that groundwater was conserved and not overexploited, so as to enable its transfer from wellfields in the A-K basin to supplement surface water sources, which due to the consecutive failure of the monsoon rains were inadequate to meet the requirements for drinking and other domestic purposes of the people in the city of Chennai. The Act considered this to be necessary in the public interest. The Act required any person wishing to abstract, use or transport groundwater for any purpose other than domestic purposes, to obtain a licence. The granting of a licence must take account of a number of factors, including the purpose or purposes for which groundwater is to be used, the existence of other competitive users, availability, effect on other sources, and pollution control. The Act required a register of existing wells and of the use of groundwater to be prepared, though a time period for this was not stipulated. The Act was a progressive first step in the management of groundwater resources, notably because it introduced an abstraction licence system that deviated from the traditional principle of the absolute ownership doctrine. For a number of reasons, the Act was not enforced. First, a 'top-down' approach was adopted that did not ensure that the interests of all stakeholders were protected. Second, there was no consultation with farmers, who represented the main stakeholder group. Third, there was no provision for compensation to farmers for reducing their agricultural activities. More generally, the focus of the Act was on conserving the groundwater to supply Chennai whilst ignoring the needs of the farmers who felt their livelihood to be threatened. Essentially, the Act was too ambitious and lacked the specifics of providing the necessary institutional, technical and financial support to enable its application.

Socio-economic and cultural aspects and stakeholders' perceptions

In the past, there was scant attention paid to the views and concerns of farmers and other stakeholders. Probably this is was one of the contributory reasons for the failure to enforce the proposed changes of the 1987 Groundwater Act. In view of this, the study undertook a sociological survey of the A-K basin farmers and other stakeholders. In addition, there was a series of workshops during which the reasons for the continued depletion of groundwater and the concepts of groundwater rights and transferable groundwater rights were explained using simple illustrations. From responses to questionnaires and discussions in workshops, it has been possible to better understand the beliefs and fears of local farmers and their resistance to reforms.

The A-K basin is primarily agricultural with a few small towns of less than 100,000 people in total, and about 300 villages, of generally 1,000–2,000 inhabitants. There has been a trend of urbanisation in recent years and industrial development close to Chennai. The traditional caste system applies with dominant farming communities, who own the land and have a considerable say in village affairs, and the Scheduled Castes or *dalits*. Most of the latter are landless labourers. The main villages are inhabited by the more socially forward castes, who also represent the dominant farming communities, while the backward castes mostly tend to live in hamlets, and the *dalits* in separate colonies. Superimposed on the traditional caste system, are political ideologies represented by political parties of the left and right. Therefore, responses to external messages and community actions are determined by a multiplicity of interlinked factors, such as caste, ethnic background, political allegiances, economic position etc.

At the time of the study, the area under cultivation was approximately 135,000 ha, and the total number of farms approximately 216,000. About 84 per cent of the farms are less than 1 ha in size, about 10 per cent 1–2 ha, and just under 6 per cent 2–5 ha. Farms larger than 5 ha amount to only 0.16 per cent of the total. The estimated number of farms greater than 1 ha in size is approximately 35,000, and the total number of potential groundwater right holders 40,000. Extensive irrigated agriculture, mainly rice paddy, has been taking place for a number of years, but this has not always been the case. About 60 years ago farmers used to raise just one paddy crop, and irrigation was from surface water impounded in shallow reservoirs (tanks) and from non-mechanised shallow hand-dug wells. The cultivated area was limited, restricted to the *ayacat* (command area) of the tanks or what could be irrigated from wells. With the introduction of electric pumps, free electricity and machinery capable of constructing deep boreholes, more areas were brought under cultivation using groundwater and paddy was raised by most farmers during all three crop seasons. Surface water structures were increasingly abandoned. Rice cultivation became a strong tradition, which was strengthened by the popular perception in the local area that it is noble to grow one's own rice. Failing to do so became a sign of reduced status of the person or the family concerned. There are also village rituals associated with agriculture and there is the belief that village deities will bless the village with enough water for cultivation provided sacrifices known as *kavu* are given to the village deities during festivals. This makes it easier to understand why farmers have been reluctant to accept that the depletion of groundwater has been largely of their own making, due to overabstraction. Sand mining of river beds has been seen by farmers as a major reason for groundwater depletion. However, hydrogeological studies suggest that this is only a local problem. Farmers feel that water is a free gift of nature and that it is their right that they should have enough of it to raise the three crops that have become customary.

Although farmers do not include the cost of water in their crop price calculations, they do attach a value to water when big farmers sell water to small farmers (for example, so many bags of rice for every acre irrigated) or when they sell water to CMWSSB. In the latter case, the negotiated price at the time was 27 rupees per hour of pumping, which corresponds to approximately 22 rupees per cubic metre. Electricity is supplied free to farmers. The price was calculated on the basis of pumping equipment and maintenance costs, labour costs and loss of revenue from crops.

Flood irrigation, the method used for paddy irrigation, consumes much water. However, as indicated above, approximately 40 per cent of the applied water returns to the water table. Generally, farmers seem to be non-responsive to the use of alternative crops or of less water-consuming irrigation methods. They cite lack of available labour (this is disputed by the labourers who complain that farmers will not use them for fear of having to raise their wages) and lack of information.

Land owners in the A-K basin are familiar with the concept of title deeds (*pattas*) for land but also other similar *pattas*, such as tree *pattas*. Farmers in general, but not older farmers, seem to accept in principle the concept of groundwater rights being allocated separately from land. They appear to accept also the definition of groundwater rights in terms of volume over time that can be varied depending on availability, providing that equity prevails. There is fear amongst small farmers who do not own pumps, however, that they may be disadvantaged. With regard to the introduction of TGR there was concern on the impact on the value of land, if the water beneath it were to belong to another person. Also, the fear that TGR was a device that would eventually lead to the privatisation of the groundwater of the A-K basin, and concern over the fate of landless labourers, who rely for their livelihood on agriculture, should most of the water be transferred out of agriculture to public supply.

In summary, the sociological studies revealed that farmers were fairly aware of the issues of groundwater depletion, but would not accept that they contributed to it. They generally considered that they were entitled to free water and that the ability to grow rice for one's home use enhanced social standing and status, and should not be denied by conveying water elsewhere, even for payment. The concept of variable volumetric water rights as a resource management tool did not meet with much resistance, but groundwater rights that could be transferred separately from land raised serious concerns about the impact on land values, the fate of landless agricultural labourers and the beginning of a process that could lead to the privatisation of the groundwater of the basin.

Proposed legal reforms and other measures

Before addressing the question of legal reforms to facilitate the introduction of TGR, it was considered important to attend to the following issues:

1   The current practice of purchase of water from farmers by the CMWSSB without the underpinning of regulatory measures. This was not considered to be a long-term viable proposition for two reasons: first, it leads to overabstraction, i.e. farmers not only abstract water to sell but also to meet their own irrigation requirements, and, second, it is socially divisive as it creates conflicts between those farmers who are able to sell water and those who cannot. This social division is compounded by the fact that overabstraction by a farmer selling water to CMWSSB results in excessive drawdown in wells of neighbouring farmers.

2   The rehabilitation and improvement of existing surface water systems. This was considered to be a good way of increasing the volume of water to agriculture, which fell from around 125 $Mm^3 a^{-1}$ in 1970–1980 to 74 $Mm^3 a^{-1}$ in 1990–2002.

A number of measures were considered but rejected, including:

1   artificial recharge which was considered unlikely to be able to augment the groundwater system in large quantities, although locally it could be useful;

2   alternative irrigation practices and crops that save water.

On the basis of farmers' responses, these were unlikely to be adopted on a reasonably large scale in the near future, for a number of reasons, including costs of establishing such systems, crop agronomy and investment risks, lack of technical and management skills to operate sophisticated systems (drip or sprinkler) and the preference for growing rice for the social and cultural reasons given above.

The study considered that only such legislative reforms that encouraged the gradual reduction of abstraction while sensitive to the concerns of all stakeholders were likely to bring about in the long term a more sustainable use of the aquifer. It was considered important that the role of the state as the trustee to the natural resources of the nation should be explicitly stated, thus giving effect to Article 39 of the Constitution and to recent decisions of the Supreme Court. Objectives of the proposed new legislation included:

1   the gradual reduction of abstractions to sustainable levels;

2   ensuring that groundwater rights were secure;

3   a recognition of the ecological, economic and social value of water;
4   the safeguarding of the interests and livelihoods of the poor and disadvantaged.

Groundwater rights were to be defined volumetrically, the initial volumes to be assigned on the basis of prior use. Volumetric rights would be variable, and varied annually on the basis of hydrological conditions. The ultimate aim would be for the aquifer to achieve its sustainable yield. To this end, it was proposed that the A-K basin should be divided into a number of extraction management zones (EMZ), each assigned a sustainable yield value, representing the desired upper limit of abstraction. A continuous process of assessment of the state of saturation of the aquifer based on groundwater level time sequences from strategically placed 'trigger station' monitoring wells would enable the Aquifer Agency to adjust the volumes of groundwater rights.

The total volume of assigned groundwater rights in each EMZ should eventually not exceed the sustainable yield limit. This may take many years to achieve. Groundwater rights may be cancelled where it was considered necessary for the public interest, subject to provision of water from an alternative source or the payment of full and fair compensation, or if the water had not been used for more than three years without good reason.

It was proposed that an institutional framework should be set up comprising an Aquifer Agency responsible for the overall management of the A-K basin aquifer. Reforms in the form of TGR should be introduced but subject to a number of conditions, namely:

1   groundwater rights for domestic purposes could not be transferred separately from the parcel of land on which they occurred;
2   groundwater rights could not be transferred from one district (*panchayat*) to another;
3   approval from the Aquifer Agency would be required for a groundwater right to be used at a different abstraction point.

Factors that would determine whether or not a transfer should take place would be: water availability, potential impact on existing users in the vicinity, and type of water use. Where water use would change from agriculture to another use, the volume of the transferred water right would automatically decrease to 60 per cent of the original quantity, this in order to reflect the loss to groundwater from irrigation returns. The proposed legislation also allowed for the purchase and retirement of water rights by the Aquifer Agency or by any other person on a voluntary basis. Once, a groundwater right was retired it would not be possible for the Agency to seek to reallocate it.

As already indicated, the concept of TGR was seen by some of the stakeholders as ultimately having the (hidden) objective of privatisation of the groundwater of the basin. It was explained that the legislative proposals were primarily related to the allocation of volumetric water rights, and that their transferability was between willing buyers and sellers, on the basis of heavy regulation aimed at avoiding the undesirable consequences of monopolies, social inequity and overabstraction. Despite the explanations, the stakeholders' fears remained, and were judged to be serious enough to potentially present a significant obstacle to implementation. The proposals were therefore modified and presented as a second option for consideration. The term 'water right' and the emotive term 'transferable water right' were replaced by the term 'volumetric extraction licence'. More importantly, permanent transfers were only permitted by way of succession or to a state utility pursuant to its statutory public water supply obligations in return for equitable compensation in the form of an equivalent volumetric water supply from an alternative source or equitable financial compensation. A transfer could only proceed if the licence holder were willing to participate in the process.

The study concluded that the pursuit of the first option although closer to the kind of radical reform favoured by the World Bank, carried with it the serious risk of no reform taking place, with the aquifer continuing to be exploited at an unsustainable level and the CMWSSB wellfields remaining largely unused. The second option while achieving the main objectives would allow stakeholders to become familiar with the practical benefits of volumetric water allocations, and set in train, on the basis of experience, the possibility for other more radical changes subject to stakeholder approval.

The situation today

To date, TGR, even in a diluted form have not been introduced in the A-K groundwater basin. It appears that this has been largely due to strong concerns about the principle of trading of groundwater rights. Also, in the last few years, the monsoons have been heavy, which has reduced the urgency of the situation.

Summary

The introduction of TGR is not just a matter of economic reform but a multifaceted process that touches diverse aspects of human existence and man's social and physical environment. The proposed generic TGR model should assist the process. A systematic approach is important as it allows diverse factors to be objectively examined. This is particularly relevant in many of the developing countries in which irrigation from groundwater has led to the depletion of many aquifers, but also in which vulnerable communities may be easily disadvantaged.

References

Charalambous, A. N. and Garratt, P. (2009) Recharge-abstraction relationships and sustainable yield in the Arani-Kortalaiyar groundwater basin, India. *Quarterly Journal of Engineering Geology and Hydrogeology*, 42(1): 39–50.

Scott Wilson Piésold in association with Hydrogeological Services International Limited, Four Seasons Marketing PVT (Delhi), Centre for Development Research and Training

(Chennai) and Overseas Development Group East Anglia (2004) *Second Chennai Water Supply Project. Phase 1. The reassessment of groundwater potential and transferable water rights in the A-K Basin.* Unpublished report to Chennai Metropolitan Water Supply and Sewerage Board, India.

Scott Wilson Piésold in association with Hydrogeological Services International Limited, Four Seasons Marketing PVT (Delhi), Centre for Development Research and Training (Chennai) and Overseas Development Group East Anglia (2006) *Second Chennai Water Supply Project. Phase 1. The reassessment of groundwater potential and transferable water rights in the A-K Basin.* Unpublished report to Chennai Metropolitan Water Supply and Sewerage Board, India.

UNDP (United Nations Development Programme) (1985) *Hydrogeology and artificial recharge studies*, Madras. Phase 1, Project DP/UN/IND-78-029/2. Mid Term Report. New York: United Nations.

UNDP (United Nations Development Programme) (1987) *Hydrogeology and artificial recharge studies*, Final Technical Report. Madras. Phase 1, Project DP/UN/IND-78-029/2. New York: United Nations.

8 Overview and conclusions

The increasing demand for water has meant that many of the world's aquifers have been over-exploited, mostly as a result of abstractions for irrigation. In South Asia, irrigation, mainly for paddy rice, represents some 75 per cent of total abstraction, whilst in some African and South American countries, mining activities in desert areas have been relying on groundwater. In the last 20–30 years expanding cities have added to the demand for potable water supplies. The consequences have been falling groundwater levels, depletion of aquifers, contamination of coastal aquifers and reductions in river flows. There have also been environmental effects on the ecology of wetlands and rivers and social effects in countries with pastoral or indigenous populations. In some cities excess groundwater withdrawals have caused ground subsidence. On the positive side, the greater use of groundwater has had many benefits. Agriculture has flourished, often encouraged by government in the form of energy subsidies, which enabled poor farming communities to feed themselves. Mining has helped the economies of countries to develop and clean groundwater has improved the health and wellbeing of rural communities as well as urban populations. If groundwater is to continue to serve both humans and the environment, it must be used sustainably. This can be achieved in a variety of ways. Some are technical, involving more efficient irrigation methods, reduction of leakage in water supply distribution systems or storage of surplus water in aquifers. Others are based on economic approaches, such as water rights trading, and others on legal measures controlling abstraction from vulnerable aquifers. Even with all these measures in place, local water resources are at times not sufficient to meet all competing demands and the answer may lie in seeking alternative sources. Bulk water transfers from areas of plenty to areas of scarcity is one solution. Such transfers are usually large engineering undertakings conveying water across different states or watersheds, often over large distances through difficult terrain. Their implementation requires sensitivity to the environment and the impacts on local inhabitants both in the short and long term. More exotic solutions are the desalination of seawater or the treatment of contaminated and brackish groundwater. They are both energy intensive and costly but for islands and dry areas may be the only alternative.

Much scientific progress has been made during the last century in understanding the principles that govern groundwater. As a result, groundwater has become less mysterious. Groundwater is hidden underground moving slowly in a diffuse manner through the interstices of rocks, known as aquifers. Observation and theory have shown that the effects of pumping groundwater from an aquifer can spread away from the point of abstraction and affect other users, springs, river flows and wetlands. Mathematical equations and computer models can be used to quantitatively estimate these effects. The great majority of aquifers are recharged by the infiltration of rainfall. In these aquifers, groundwater is renewable. In arid and semi-arid areas recharge is low, only a small percentage of rainfall, and only happens occasionally. In some deserts there has been no recharge for tens of thousands of years. Groundwater in the aquifers of these areas is of fossil origin and is not renewable. Exploitation of these aquifers is from storage reserves and a planned development is required, if their untimely depletion is to be avoided. Even rechargeable aquifers are exhaustible and their exploitation must, on average, balance recharge. Many of the world's large aquifers cross state or national boundaries. These shared or transboundary aquifers are found mainly in the Sahara Desert of North Africa, in the arid and semi-arid Middle East and in humid South America. Their exploitation poses special problems of ownership, allocation and monitoring between sharing aquifer states. International law to deal with the issues of transboundary aquifers is in the process of development.

The legal understanding of groundwater is very similar to the modern scientific understanding. The law correctly distinguishes between groundwater and subterranean streams which follow a more or less defined course. The latter are treated in the same way as surface water streams. Water law has its origins in Roman law, the civil law of the continent of Europe and the common law of England. In all three traditions, a landowner owns the groundwater beneath his land and can abstract as much water as he likes without concern as to whether his neighbour might be affected. This has come to be known as the absolute ownership doctrine or English doctrine and in Texas USA, the law of capture. It is still practised in many developing countries and in a few states of the western USA. There are two other traditional water rights doctrines, the riparian doctrine and the prior appropriation doctrine, although neither is directly related to groundwater. Like the doctrine of absolute ownership, the riparian doctrine is a land-based doctrine. It is concerned with the water rights of owners of land abutting river courses. The prior appropriation doctrine is fundamentally of an American origin. It allowed miners in the arid western USA to divert water from a water source without needing to own the land. Its significance lies in the fact that water rights could be owned independently of land ownership. The traditional doctrines lacked definition and as a result, in addition to extensive litigation, led to overabstraction and adverse impacts on the environment. Statutory permit type systems have been progressively replacing the old doctrines. In these, water rights are formally and explicitly defined, including volume and duration. The unlimited use of groundwater without regard to the impact on neighbours or the environment is no longer permitted. In modern water codes, water resources are owned or held in trust by the state and groundwater rights have become usufructuary. In essence this means that landowners may use the groundwater that occurs under their land but do not own it. The land and water link is still generally maintained. In a few countries, water rights have been assigned property rights in order to make them transferable independently of land. There has been a debate on whether water rights are property rights. One objection is that unlike land water, even slow moving groundwater, cannot be easily possessed. Nevertheless, the general consensus has been that water rights are a form of property rights, although not in the same category as land rights.

In economic terms, groundwater is probably best thought of as an open access good. Being a common pool resource, others cannot be easily excluded from using it and as a finite resource, even when renewable, its consumption leaves less for others to enjoy. Groundwater is also a merit good that has no substitute. Governments have a responsibility to ensure that people, even when they are not able to pay, cannot be denied access to it. In its natural state, groundwater is generally not considered to be a commodity or product. It may become so after it is captured and processed for the sole purpose of sale, as is the case with bottled mineral waters or waters in soft or aerated drinks.

In the sense that available groundwater resources cannot satisfy all the alternative uses of water (irrigation, potable supply, industrial use, environmental allocations, etc.) they are scarce resources, and as such assume the character of an economic good. Economic principles to manage scarce groundwater resources have been proposed by institutions such as the World Bank and the United Nations, especially for developing countries where large volumes of cheap groundwater are used for the irrigation. Neo-liberal economists have argued that in a free market, water will be used more efficiently and be able to achieve its economic value. Moreover, in an ideal market of many sellers and many buyers, supply and demand will regulate the use of groundwater to a sustainable level. In the free-market approach groundwater is essentially treated as a commodity to be bought and sold in the same way as other private goods. As groundwater is not a private good, it is subject to negative externalities. Some of these can be significant, as they are directly related to the survival of human beings, the environment and social equity. The market is generally not very good in responding to externalities and almost invariably adverse impacts require government intervention. Thus, traded groundwater values rarely reflect the total economic value (TEV). Methodologies have been developed to evaluate TEV based on willingness-to-pay (WTP) and willingness-to-accept (WTA) compensation. But these do not always provide accurate or objective values as they often reflect the individual's preferences, which may be influenced by self interest that does not always coincide with the collective benefit to society or non-human entities. Nevertheless, economic valuations and monetary values are a useful tool that provides information on how scarce resources may be efficiently managed.

Traditionally, the transfer of water rights has been achieved through the sale or transfer of land. This has been a constraint to groundwater rights trading. In transferable groundwater rights (TGR) the land–water link has been severed, thus enabling groundwater rights to be traded independently of the land parcel under which the groundwater occurs. Legislation severing the land–water link and assigning property rights to water rights has been introduced in a few countries, but, generally, not in South Asia, where groundwater is widely used in large quantities for irrigation. With the possible exception of the western USA, TGR have been a recent phenomenon, partly due to the recognition that water is a scarce resource and partly as a result of the free market ideological reforms of the 1980s and 1990s. TGR have been applied to countries with a range of regulatory regimes. In Chile, the 1981 Water Code assigned private property rights to water resources and until recently with little or no regulation. In Mexico, the 1992 National Water Law and regulations in 1998 allowed water rights to be transferred separately from land rights. However, unlike Chile, in Mexico the state has strong regulatory powers over water resources management and allocation. Australia, through various water acts and inititiatives, has been actively promoting water rights trading since 1997 although still largely under a regulatory framework. The western USA has had the longest experience following from the introduction of the prior appropriation doctrine in the nineteenth century, but here also water rights trades are subject to conditions laid down by legislation. In England and Wales, the 2003 Water Act allowed water rights trading, however, all trades must be approved by the Environment Agency. A new (2012) draft water bill seeks to facilitate further water rights trading.

Water rights transfers have been mainly in surface water, especially in hydrologically connected basins, such as the Limari basin in Chile and the southern Murray–Darling basin in Australia or Colorado and California in the western USA. In these countries, water transfers were facilitated by large government-funded engineering projects constructed during the last century and before. So far, TGR have not been used to transfer bulk quantities of groundwater from agriculture to higher value urban or industrial uses. Proposals based on studies partly funded by the World Bank in 2002–2006 to utilise TGR to transfer irrigation groundwater from the Arani-Kortalaiyar basin in south India to public water supply for Chennai have not yet been implemented, probably for fear by some that this might lead to the privatisation of water resources.

The evidence to date suggests that overall water rights trading has not contributed significantly to the sustainable management of water resources. In fact water rights trades have generally amounted to only a small percentage of total use. In Chile, where trading has been allowed to proceed without state interference, water rights have been overallocated with groundwater being overabstracted in most areas. This forced government to introduce new legislation to protect aquifers and ecological flows. A similar situation exists in the western USA where many aquifers have been depleted, despite water rights trading having taken place for many years. In Mexico groundwater continues to be overexploited with the market being unable to provide substantive solutions and, recently, resorting to water banks to activate the market. In Australia groundwater trading has been very weak, amounting to a one or two per cent of total groundwater use, and this despite considerable and costly administrative reforms over a number of years.

In England and Wales, where unlike other countries only less than 1 per cent of total use goes to irrigation, the trading of water rights has been insignificant, 0.002 per cent of total abstraction. Yet despite this, further reforms are being proposed in the hope that the market may become more responsive, including water rights buy-backs at government expense to protect the environment from overabstraction.

TGR have been useful as a means for reducing abstractions to sustainable levels mainly through buy-backs by the state. Examples are the Edwards Limestone aquifer in Texas and more recently the southern Murray–Darling basin in Australia, although here mainly in surface water. In Texas concerns over the impact of the reduction of springflows on the ecosystems due to excessive groundwater abstraction led to the enactment of the Edwards Aquifer Act in 1993. This enabled the Edwards Aquifer Authority to acquire water rights from existing users in order to reduce abstraction to the sustainable level. In the southern Murray–Darling basin, in 2009–2010 the government bought water rights from irrigators amounting to 500 Mm^3 in order to safeguard the environment.

In the late 1980s and early 1990s evaluations of the performance of water rights trading, mainly relating to Chile, were generally favourable. This was a time of euphoria that saw the free-market economies of Western Europe and the United States triumph over the centrally controlled economies of Eastern Europe, the Soviet Union and China. Thus, western nations, starting with England and Wales, went through a phase of deregulation and privatisation of state-owned industries, including the water sector. Exporting this model to developing countries, Eastern Europe and the ex-Soviet Union satellite states became almost something of a missionary crusade. In the first decade of the new century cracks began to appear. The collapse of large companies such as the water and energy giant Enron, financial crises in the banking sector that needed government intervention and increases in the prices of food and water have reinforced the belief that the private sector, which has profit as its primary objective, cannot be entirely relied upon to contain the social effects of its actions. This is probably particularly true for a resource such as water which humanity and the environment cannot do without. Developing countries have been reluctant to introduce water rights trading, which still remains a sensitive issue. The recent performance of financial markets is not likely to have allayed fears regarding the competence of the private sector and reliance on market forces.

Introducing TGR is not just a matter of achieving free market economic objectives. There are other important factors which must be taken into account, such as physical constraints, legislative aspects, political perceptions, environmental protection, social equity, and sensitivity to cultural and religious beliefs. In this book, a generic flow chart model has been presented, which may be used as a guide in the decision process to introduce TGR.

Potentially TGR offer a better alternative for the management of groundwater resources than land-based water rights systems. However, it is doubtful whether free trade, even in a perfect market, will be able to achieve groundwater resources sustainability, let alone protect the environment or safeguard the welfare and livelihood of the poor and vulnerable. Legal codes allowing water rights trading should therefore include regulatory measures that deal with potentially adverse impacts. A completely unregulated system of TGR is not recommended. TGR should probably be the final step in water resources reforms aimed at assigning an independent status to water rights. In developing countries with large rural communities that rely on groundwater for their water supply and food production, TGR may not be appropriate in the first instance. Reforms to limit abstraction may start with the introduction of formal and explicit water rights based on abstraction permit systems.

Index

Printed and bound by CPI Group (UK) Ltd, Croydon, CR0 4YY

22/10/2024

01777623-0012